Quantum Mechanics

Springer
*Berlin
Heidelberg
New York
Hong Kong
London
Milan
Paris
Tokyo*

A primeval representation of the hydrogen atom

This beautiful mandala is displayed at the temple court of Paro Dzong, the monumental fortress of Western Bhutan [1]. It may be a primeval representation of the hydrogen atom: the outer red circle conveys a meaning of strength, which may correspond to the electron binding energy. The inner and spherical nucleus is surrounded by large, osculating circles that represent the motion of the electron: the circles do not only occupy a finite region of space (as in Fig. 6.4), but are also associated with trajectories of different energies (colours) and/or with radiation transitions of different colours (wavelengths). At the center, within the nucleus, there are three quarks.

Daniel R. Bes

Quantum Mechanics

A Modern and Concise
Introductory Course

With 52 Figures

 Springer

Prof. Daniel R. Bes
Physics Department, CNEA
Av. del Libertador 8250
Buenos Aires 1429
Argentina
e-mail: bes@tandar.cnea.gov.ar

and

Physics Department
Universidad Favaloro
Belgrano 1723
Buenos Aires 1093
Argentina
e-mail: bes@favaloro.edu.ar

Library of Congress Cataloging-in-Publication Data.
Bes, Daniel R.
Quantum mechanics: a modern and concise introduction/Daniel R. Bes.
p.cm. Includes bibliographical references and index.
ISBN 3-540-20365-6 (acid-free paper)
1. Quantum theory. I. Title.
QC174.45B43 2004   530.12–dc22   2004040688

ISSN 1439-2674

ISBN 3-540-20365-6 Springer-Verlag Berlin Heidelberg New York

This work is subject to copyright. All rights are reserved, whether the whole or part of the material is concerned, specifically the rights of translation, reprinting, reuse of illustrations, recitation, broadcasting, reproduction on microfilm or in any other way, and storage in data banks. Duplication of this publication or parts thereof is permitted only under the provisions of the German Copyright Law of September 9, 1965, in its current version, and permission for use must always be obtained from Springer-Verlag. Violations are liable for prosecution under the German Copyright Law.

Springer-Verlag is a part of Springer Science+Business Media

springeronline.com

© Springer-Verlag Berlin Heidelberg 2004
Printed in Germany

The use of general descriptive names, registered names, trademarks, etc. in this publication does not imply, even in the absence of a specific statement, that such names are exempt from the relevant protective laws and regulations and therefore free for general use.

Typesetting: Data prepared by the author using a Springer TEX macro package
Final processing by Frank Herweg, Leutershausen
Cover design: *design & production* GmbH, Heidelberg

Printed on acid-free paper   SPIN 10959862   57/3141/tr   5 4 3 2 1 0

Foreword

Quantum mechanics is undergoing a revolution. Not that its substance is changing, but two major developments are placing it in the focus of renewed attention, both within the physics community and among the scientifically interested public. First, wonderfully clever table-top experiments involving the manipulation of single photons, atomic particles and molecules are revealing in an ever-more convincing manner theoretically predicted facts about the counterintuitive and sometimes 'spooky' behavior of quantum systems. Second, the prospect of building quantum computers with enormously increased capacity of information-processing is fast approaching reality. Both developments demand more and better training in quantum mechanics at the universities, with emphasis on a clear and solid *understanding* of the subject.

Cookbook-style learning of quantum mechanics, in which equations and methods for their solution are memorized rather than understood, may help the students to solve some standard problems and pass multiple-choice tests, but it will not enable them to break new ground in real life as physicists. On the other hand, some 'Mickey Mouse courses' on quantum mechanics for engineers, biologists and computer analysts may give an idea of what this discipline is about, but too often the student ends up with an incorrect picture or, at best, a bunch of uncritical, blind beliefs. The present book represents a fresh start toward helping achieve a deep understanding of the subject. It presents the material with utmost rigor and will require from the students ironclad, old-fashioned discipline in their study.

Too frequently, in today's universities, we hear the demand that the courses offered be 'entertaining', in response to which some departmental brochures declare that 'physics is fun'! Studying physics requires many hours of hard work, deep concentration, long discussions with buddies, periodic consultation with faculty, and tough self-discipline. But it can, and should, become a passion: the *passion* to achieve a deep understanding of how Nature works. This understanding usually comes in discrete steps, and students will experience such a step-wise mode of progress as they work diligently through the present book. The satisfaction of having successfully mastered each step will then indeed feel very rewarding!

The 'amount of information per unit surface' of text is very high in this book – its pages cover *all* the important aspects of present-day quantum mechanics, from the one-dimensional harmonic oscillator to teleportation. Starting from a few basic principles and concentrating on the fundamental behavior of systems (particles) with only a few degrees of freedom (low-dimension Hilbert space), allows the author to plunge right into the core of quantum mechanics. It also makes it possible to introduce *first* the Heisenberg matrix approach-in my opinion, a pedagogically rewarding method that helps sharpen the mental and mathematical tools needed in this discipline right at the beginning. For instance, to solve the quantization of the harmonic oscillator *without* the recourse of a differential equation is illuminating and teaches dexterity in handling the vector and matrix representation of states and operators, respectively.

Daniel Bes is a child of the Copenhagen school. Honed in one of the cradles of quantum mechanics by Åge Bohr, son of the great master, and by Ben Mottelson, he developed an unusually acute understanding of the subject, which after years of maturing he now has projected into a book. The emphasis given throughout the text to the fundamental role and meaning of the measurement process, and its intimate connection to Heisenberg's principle of uncertainty and non-commutativity, will help the student overcome the initial reaction to the counterintuitive aspects of quantum mechanics and to better comprehend the physical meaning and properties of Schrödinger's wave function. The human brain is an eminently classical system (albeit the most complex one in the Universe as we know it), whose phylo- and onto-genetic evolution were driven by *classical* physical and informational interactions between organism and the environment. The neural representations of environmental and ontological configurations and events, too, involve eminently classical entities. It is therefore only natural that when this classical brain looks into the microscopic domain using human-designed instruments which must translate quantum happenings into classical, macroscopically observable effects, strange things with unfamiliar behavior may be seen! And it is only natural that, thus, the observer's intentions and his instruments cannot be left outside the framework of quantum physics! Bes' book helps to recognize, understand and accept quantum 'paradoxes' not as such but as the facts of 'Nature under observation'. Once this acceptance has settled in the mind, the student will have developed a true intuition or, as the author likes to call it, a 'feeling' for quantum mechanics.

Chapter 2 contains the real foundation on which quantum mechanics is built; it thus deserves, in my opinion, repeated readings-not just at the beginning, but after each subsequent chapter. With the exception of the discussion of two additional principles, the rest of the book describes the mathematical formulations of quantum mechanics (both the Heisenberg matrix mechanics, most suitable for the treatment of low-dimension state vectors, and Schrdinger's wave mechanics for continuous variables) as well as many applications. The examples cover a wide variety of topics, from the simple harmonic oscillator mentioned above, to subjects in condensed matter physics, nuclear physics, electrodynamics and quantum computing. It is interesting to note, regarding the latter, that the concept of qubit appears in a most natural way in the middle of the book, almost in passing, well before the essence of quantum computing is discussed towards the end. Of particular help are the carefully thought-out problems at the end of each chapter, as well as the occasional listings of 'common misconceptions'. A most welcome touch is the inclusion of a final chapter on the history of theoretical quantum mechanics – indeed, it is regrettable that so little attention is given to it in university physics curricula: much additional understanding can be gained from learning how ideas have matured (or failed) during the historical development of a given discipline!

Let me conclude with a personal note. I have known Daniel Bes for over sixty years. As a matter of fact, I had known of him even before we met in person: our fathers were 'commuter-train acquaintances' in Buenos Aires, and both served on the PTA of the primary school that Daniel and I attended (in different grades). Daniel and I were physics students at the University of Buenos Aires (again, at different levels), then on the physics faculty, and years later, visiting staff members of Los Alamos. We always were friends, but we never worked together – Daniel was a theoretician almost from the beginning, whereas I started as a cosmic-ray and elementary-particle experimentalist (see Fig. 2.4!). It gives me a particular pleasure that now, after so many years and despite residing at opposite ends of the American continent, we have become professionally 'entangled' through this wonderful textbook!

University of Alaska-Fairbanks *Juan G. Roederer*
January 2004 Professor of Physics Emeritus

Preface

This text follows the tradition of starting an exposition of quantum mechanics with the presentation of the basic principles. This approach is logically pleasing and it is easy for students to comprehend. Paul Dirac, Richard Feynman and, more recently, Julian Schwinger, have all written texts which are epitomies of this approach.

However, up to now, texts adopting this line of presentation cannot be considered as introductory courses. The aim of the present book is to make this approach to quantum mechanics available to undergraduate and first year graduate students, or their equivalent.

A systematic dual presentation of both the Heisenberg and Schrödinger procedures is made, with the purpose of getting as quickly as possible to concrete and modern illustrations. As befits an introductory text, the traditional material on one- and three-dimensional problems, many-body systems, approximation methods and time-dependence is included. In addition, modern examples are also presented. For instance, the ever-useful harmonic oscillator is applied not only to the description of molecules, nuclei and the radiation field, but also to recent experimental findings, like Bose–Einstein condensation and the integer quantum Hall effect.

This approach also pays dividends through the natural appearance of the most quantum of all operators: the spin. In addition to its intrinsic conceptual value, spin allows us to simplify discussions on fundamental quantum phenomena like interference and entanglement; on time-dependence (as in nuclear magnetic resonance); and on applications of quantum mechanics in the field of quantum information.

This text permits two different readings: one is to take the shorter path to operating with the formalism within some particular branch of physics (solid state, molecular, atomic, nuclear, etc.) by progressing straightforwardly from Chap. 2 through to Chap. 9. The other option, for computer scientists and for those readers more interested in applications like cryptography and teleportation, is to skip Chaps. 4, 6, 7 and 8, in order to get as soon as possible to Chap. 10, which starts with a presentation of the concept of entanglement. Chapter 11 is devoted to a further discussion of measurements and interpretations in quantum mechanics. A brief history of quantum mechanics is presented in order to acquaint the newcomer with the development of one of the most spectacular adventures of the human mind to date (Chap. 12). It intends also to convey the feeling that, far from being finished, this enterprise is continually being updated.

Sections labeled by an asterisk include either the mathematical background of material that has been previously presented, or display a somewhat more advanced degree of difficulty. These last ones may be left for a second reading.

Any presentation of material from many different branches of physics requires the assistance of experts in the respective fields. I am most indebted for corrections and/or suggestions to my colleagues and friends Ben Bayman, Horacio Ceva, Osvaldo Civitarese, Roberto Liotta, Juan Pablo Paz, Alberto Pignotti, Juan Roederer, Marcos Saraceno, Norberto Scoccola and Guillermo Zemba. However, none of the remaining mistakes can be attributed to them. Civitarese and Scoccola also helped me a great deal with the manuscript.

Questions (and the dreaded absence of them) from students in courses given at Universidad Favaloro (UF) and Universidad de Buenos Aires (UBA) were another source of improvements. Sharing teaching duties with Guido Berlin, Cecilia López and Darío Mitnik at UBA was a plus. The interest of Ricardo Pichel (UF) is fully appreciated.

Thanks for corrections to my English are due to Peter Willshaw. The help of Martin Mizrahi and Ruben Weht with drawing the figures is gratefully acknowledged. Raul Bava called my attention to the mandala on p. II.

I would like to express my appreciation to Arturo López Dávalos for putting me in contact with Springer-Verlag and to Angela Lahee and Petra Treiber of Springer-Verlag for their help.

My training as a physicist owes very much to Åge Bohr and B. Mottelson of Niels Bohr Institutet and NORDITA (Copenhagen). During the fifties Niels Bohr, in his long-standing tradition of receiving visitors from all over the world, used his institute as an open place where physicists from East and West could work together and understand each other. From 1956 to 1959, I was there as a young representative of the South. My wife and I met Margrethe and Niels Bohr at their home in Carlsberg. I remember gathering there with other visitors listening to Bohr's profound and humorous conversation. He was a kind of father figure, complete with a pipe that would go out innumerable times while he was talking. Years later I became a frequent visitor to the Danish institute, but after 1962 Bohr was no longer there.

My wife Gladys carried the greatest burden while I was writing this book. It must have been difficult to be married to a man who was mentally absent for the better part of almost two years. I owe her much more than a mere acknowledgement, because she never gave up in her attempts to change this situation (as she never did on many other occasions in our life together). My three sons, David, Martin and Juan have been a permanent source of strength and help. They were able to convey their encouragement even from distant places. This is also true of Leo, Flavia and Elena, and of our two granddaughters, Carla and Lara.

My dog Mateo helped me with his demands for a walk whenever I spent too many hours sitting in front of the monitor. He does not care about Schrödinger's cats.

University of Alaska-Fairbanks *Daniel R. Bes*
January 2004 Professor of Physics Emeritus

Contents

1 **Introduction** 1

2 **Principles of Quantum Mechanics** 5
   2.1 Classical Physics 5
   2.2 Hilbert Spaces and Operators 6
   2.3 Basic Principles of Quantum Mechanics 9
   2.4 Some Consequences of the Basic Principles 12
   2.5 Measurement Process 15
      2.5.1 The Concept of Measurement 15
      2.5.2 Quantum Measurements 16
   2.6 Commutation Relations and the Uncertainty Principle 17
   2.7* Some Properties of Hermitian Operators 20
   2.8* Notions of Probability Theory 21
   Problems 22

3 **The Heisenberg Realization of Quantum Mechanics** 25
   3.1 Matrix Formalism 25
      3.1.1 A Realization of the Hilbert Space 25
      3.1.2 The Solution of the Eigenvalue Equation 26
      3.1.3 Transformation Between Sets of Basis States 27
      3.1.4 Application to 2×2 Matrices 28
   3.2 Harmonic Oscillator 31
      3.2.1 Solution of the Eigenvalue Equation 32
      3.2.2 Some Properties of the Solution 34
   Problems 35

4 **The Schrödinger Realization of Quantum Mechanics** 39
   4.1 Time-Independent Schrödinger Equation 39
      4.1.1 Probabilistic Interpretation of Wave Functions 41
   4.2 The Harmonic Oscillator Revisited 43
      4.2.1 Solution of the Schrödinger Equation 43
      4.2.2 Spatial Features of the Solutions 44
   4.3 Free Particle 47
   4.4 Infinite Square Well Potential. Electron Gas 49
   4.5 One-Dimensional Unbound Problems 51

		4.5.1	One-Step Potential...............................	51
		4.5.2	Square Barrier	54
	4.6*	Finite Square Well Potential		55
	4.7*	Band Structure of Crystals.............................		57
	Problems			60

5 Angular Momentum 63
5.1 Eigenvalues and Eigenstates............................ 63
5.1.1 Matrix Treatment 63
5.1.2 Treatment Using Position Wave Functions 65
5.2 Spin 67
5.2.1 Stern–Gerlach Experiment 67
5.2.2 Spin Formalism................................ 68
5.3 Addition of Angular Momenta 70
5.4* Details of Matrix Treatment........................... 72
5.5* Details of the Treatment of Orbital Angular Momentum 73
5.6* Coupling with Spin $s = 1/2$ 75
Problems 75

6 Three-Dimensional Hamiltonian Problems 79
6.1 Central Potentials..................................... 79
6.1.1 Coulomb and Harmonic Oscillator Potentials 80
6.2 Spin–Orbit Interaction 81
6.3 Some Elements of Scattering Theory 83
6.3.1 Boundary Conditions............................ 83
6.3.2 Expansion in Partial Waves 84
6.3.3 Cross-Sections................................. 85
6.4* Solutions to the Coulomb and Oscillator Potentials......... 87
6.5* Some Properties of Spherical Bessel Functions 90
Problems 91

7 Many-Body Problems 95
7.1 The Pauli Principle 95
7.2 Two-Electron Problems................................. 98
7.3 Periodic Tables 99
7.4 Motion of Electrons in Solids 103
7.4.1 Electron Gas 103
7.4.2* Band Structure of Crystals....................... 105
7.5* Occupation Number Representation 106
7.6* Quantum Statistics 107
7.7* Bose–Einstein Condensation............................ 109
7.8* Quantum Hall Effects 111
7.8.1* Integer Quantum Hall Effect 112
7.8.2* Fractional Quantum Hall Effect 115
Problems 116

8 Approximate Solutions to Quantum Problems 119
- 8.1 Perturbation Theory 119
- 8.2 Variational Procedure 121
- 8.3 Ground State of the He Atom 122
- 8.4 Molecules 123
  - 8.4.1 Intrinsic Motion. Covalent Binding 123
  - 8.4.2 Vibrational and Rotational Motions 125
- 8.5 Approximate Matrix Diagonalizations 128
  - 8.5.1* Approximate Treatment of Periodic Potentials 128
- 8.6* Matrix Elements
  Involving the Inverse of the Interparticle Distance 130
- 8.7* Quantization with Constraints 131
  - 8.7.1* Constraints 131
  - 8.7.2* Outline of the BRST Solution 133
- Problems 133

9 Time-Dependence in Quantum Mechanics 137
- 9.1 The Time Principle 137
- 9.2 Time-Dependence of Spin States 138
  - 9.2.1 Larmor Precession 138
  - 9.2.2 Magnetic Resonance 139
- 9.3 Sudden Change in the Hamiltonian 141
- 9.4 Time-Dependent Perturbation Theory 141
  - 9.4.1 Transition Amplitudes 142
  - 9.4.2 Constant-in-Time Perturbation 143
- 9.5 Quantum Electrodynamics for Newcomers 144
  - 9.5.1 Classical Description of the Radiation Field 145
  - 9.5.2 Quantization of the Radiation Field 146
  - 9.5.3 Interaction of Light with Particles 147
  - 9.5.4 Emission and Absorption of Radiation 148
  - 9.5.5 Selection Rules 149
  - 9.5.6 Mean Lifetime
    and Energy–Time Uncertainty Relation 150
- Problems 151

10 Entanglement
and Some Recent Applications of Quantum Mechanics 153
- 10.1 Entanglement 153
- 10.2 Quantum Cryptography 154
- 10.3 Quantum Teleportation 156
- 10.4 Quantum Computation 157
  - 10.4.1 Conceptual Framework 157
  - 10.4.2 Quantum Gates 158
  - 10.4.3 Factorization 160
- 10.5* Bell States 162

        10.6* No-Cloning Theorem 163
        10.7* Manipulations with Qubits 164
        Problems 165

**11 Measurements and Alternative Interpretations
    of Quantum Mechanics. Decoherence** 167
    11.1 Measurements and Alternative Interpretations
         of Quantum Mechanics 167
    11.2 Decoherence 169

12 A Brief History of Quantum Mechanics 173
    12.1 Social Context 173
    12.2 Old Quantum Theory ($1900 \leq t \leq 1925$) 174
    12.3 Quantum Mechanics ($1925 \leq t \leq 1928$) 177
    12.4 Philosophical Aspects 181
         12.4.1 Complementarity Principle 181
         12.4.2 Bohr and Einstein 181
         12.4.3 Recent History 183

13 Solutions to Problems and Physical Constants 185
    13.1 Solutions to Problems 185
    13.2 Physical Units and Constants 192

References 195

Index 199

1 Introduction

The construction of classical physics started at the beginning of the seventeenth century. By the end of the nineteenth century, the building appeared to have been completed: the mechanics of Galileo Galilei and of Isaac Newton, the electromagnetism of Michael Faraday and James Maxwell, and the thermodynamics of Ludwig Boltzmann and Hermann Helmholtz were by then well established, from both the theoretical and the experimental points of view. The edifice was truly completed when Albert Einstein developed the special and the general theories of relativity, in the early twentieth century.

Classical physics deals with the trajectory of particles (falling bodies, motion of planets around the sun) or with the propagation of waves (light waves, sound waves). The construction of this edifice required intuition to be abandoned in favor of a formalism, i.e., a precise treatment that predicts the evolution of the world through mathematical equations. All the formalisms applied in classical physics have a deterministic character. The existence of a physical reality, independent of the observer, is an implicitly accepted dogma.

Cracks in this conception appeared around the beginning of the last century. Light waves not only appeared to be absorbed and emitted in lumps (black-body radiation) [2], but turned out to behave completely like particles (photoelectric and Compton effects [3, 4]). Experiments with electrons displayed diffraction patterns that had up to then been seen as characteristic of waves [5]. However, the most disturbing discovery was that an atom consists of a positively charged, small, heavy nucleus, surrounded by negatively charged, light electrons [6]. According to classical physics, matter should collapse in a fraction of a second! Nor was it understood why atoms emitted light of certain wavelengths only, similar to an organ pipe that produces sounds at certain well-defined frequencies [7].

In 1913 Niels Bohr was able to explain both the stability of the hydrogen atom and the existence of discrete energy levels by means of a partial rejection of classical mechanics and electromagnetism [8]. However, this model was largely a patch. Bohr himself assumed the role of leader in the quest for an adequate formalism. In 1925 Werner Heisenberg alone [9] and, subsequently, in collaboration with Max Born and Pascual Jordan [10], developed a quantum mechanical formalism in which the classical variables of position and momentum were represented by (non-commuting) matrices. Also in 1925, Dirac introduced the idea that physical quantities are represented by operators (of which Heisenberg's matrices are just one representation) and physical states by vectors in abstract Hilbert spaces [11]. In 1926 Erwin Schrödinger produced the vectors formalism of quantum mechanics, an alternative approach based on the differential equations that bear his name [12].

There also exist other formulations of quantum mechanics, which are not covered in this book, most notably Dirac's equation including relativistic effects [13], and Feynman's summation over paths [14].

Since the introduction of two different formalisms is at least twice as difficult as the presentation of a single one, a relatively large number of undergraduate quantum mechanics textbooks confine themselves to a discussion of the Schrödinger formulation. In agreement with Schwinger's opinion[1] [15], the present author contends that such a presentation of quantum mechanics is conceptually misleading, since it leads to the impression that quantum mechanics is another branch of classical wave physics. It is not.

However, in order to present both formalisms, we must bear in mind that the precision (π) achieved in the presentation of a given subject, is limited by the amount of indeterminacy or uncertainty ($\Delta\pi$) inherent in any message. On the other hand, the simplicity or clarity (σ) of any exposition is restricted by the amount of detail ($\Delta\sigma$) that must be given in order to make the message understandable. The uncertainty of a presentation may be reduced by increasing the amount of detail, and vice versa. Bohr used to say that accuracy and clarity were complementary concepts (Sect. 12.4.1). Thus a short statement can never be precise. We may go further and state that the product of indeterminacy times the amount of detail is always larger than a constant k ($\Delta\pi \times \Delta\sigma \geq k$). The quality of textbooks should be measured by how close this product is to k, rather than by their (isolated) clarity or completeness.

There are several excellent texts of quantum mechanics covering both matrix and differential formulations. Their indeterminacy is very small. However, their use is practically limited to graduate students, because of the amount of material included. On the contrary, typical undergraduate courses tend to go into much less detail, and thereby increase the indeterminacy of their content. The present introduction to quantum mechanics reduces this common indeterminacy through the inclusion of both Heisenberg and Schrödinger formulations on an approximately equal footing. We have attempted to keep the consequent amount of necessary additional information to a minimum, since these notes are supposed to constitute an introductory course. It is up to the reader to judge how closely we have been able to approach the value k. If we have achieved our aim, a more rigorous and sufficiently simple presentation of quantum mechanics will be available to undergraduate students.

---

[1] "I have never thought that this simple wave approach [de Broglie waves and the Schrödinger equation] was acceptable as a general basis for the whole subject."

This book should be accessible to students who are reasonably proficient in linear algebra, calculus, classical mechanics and electromagnetism. Previous exposure to other mathematical and/or physics courses constitutes an advantage, but is by no means a sine qua non.

The reader will be confronted in Chap. 2 with a condensed and somewhat abstract presentation of Hilbert spaces and Hermitian operators. This early presentation implies the risk that he/she might receive the (erroneous) impression that the book is mathematically oriented, and/or that he/she will be taught mathematics instead of physics. However, Sects. 2.2, 2.7* and 2.8* include practically all the mathematical tools that are used in the text (outside of elementary linear algebra and calculus, both being prerequisites). Consistently with this 'physics' approach, the results are generally starkly presented, with few detailed derivations. (These have frequently been put in chapters labeled with an asterisk.) It is the author's contention that these derivations do not significantly contribute to filling the gap between just recognising quantum mechanical expressions and learning how to 'do' and 'feel' quantum mechanics. This last process is greatly facilitated by solving the problems at the end of each chapter (with answers provided at the end of the book). The instructor as an 'answerer' and 'motivator' of students' questions, and not merely as a 'problem solver on the blackboard', is an important catalyst in the process of filling the above-mentioned gap.

2 Principles of Quantum Mechanics

Very few introductions to quantum mechanics present an accurate overview of the subject.[1] This limitation does not constitute a great drawback, since quantum mechanics is ultimately justified by its internal consistency and power of prediction. However, in order to avoid some frequent shortcomings that exist in many presentations,[2] this introduction describes first the mathematical tools used in the formulation of quantum mechanics and then the connections between the physical world and mathematical formalism. Such links constitute the fundamental principles of quantum mechanics. They are valid for every specific realization of these principles. Subsequently, their most immediate consequences are presented: the quantum process of measurement and relations of uncertainty.

2.1 Classical Physics

If our vision of a moving object is interrupted by a large billboard, and is resumed after the reappearance of the object, we naturally assume that it has traveled all the way behind the billboard (Fig. 2.1). This is implied in the notion of physical reality, one of the postulates in the famous EPR paradox written by Einstein in collaboration with Boris Podolsky and Nathan Rosen [19]: "If, without in any way disturbing a system, we can predict with certainty the value of a physical quantity, then there is an element of physical reality corresponding to this physical quantity."

---

[1] However, excellent presentations can be found in [16], which remains a cornerstone on the subject, and in [15], [17] and [18]. On the other hand, none of these books is an introduction.

[2] In most presentations it is assumed that the solution of any wave equation for a free particle is the plane wave $\exp[i(kx - \omega t)]$. Subsequently, the operators corresponding to the momentum and energy are manipulated in order to obtain an equation yielding the plane wave as the solution. This procedure is not very satisfactory because: (i) plane waves display some difficulties as wave functions, not being square integrable; (ii) quantum mechanics appears to be based on arguments that are only valid within a differential formulation; (iii) it leads to the misconception that the position wave function is the only way to describe quantum states.

Fig. 2.1. The trajectory of the car behind the billboard as an element of physical reality

This classical framework relies on the acceptance of some preconceptions, most notably the existence of the continuous functions of time called trajectories $\boldsymbol{x}(t)$ [or $\boldsymbol{x}(t)$ and $\boldsymbol{p}(t)$, where $\boldsymbol{p}$ is the momentum of the particle]. The concept of trajectory provides an important link between the physical world and its mathematical description. For instance, it allows us to formulate Newton's second law:

$$\boldsymbol{F} = \frac{\mathrm{d}\boldsymbol{p}}{\mathrm{d}t} \,. \tag{2.1}$$

This equation of motion predicts the evolution of the system in a continuous and deterministic way. Since this is a second order equation, the state of a system is specified if the position and momentum of each particle are known.

Maxwell's theory of electromagnetism is also part of classical physics. It is characterized in terms of fields, which must be specified at every point in space and time. Unlike particles, fields can be made as small as desired. Electromagnetism is also a deterministic theory.

Essential assumptions in classical physics about both particles and fields are:

- the possibility of non-disturbing measurements,
- there is no limit to the accuracy of the values assigned to physical properties.

In fact, there is no distinction between physical properties and the numerical values they assume. Schwinger characterizes classical physics as [15], p. 11: "the idealization of non-disturbing measurements and the corresponding foundations of the mathematical representation, the consequent identification of physical properties with numbers, because nothing stands in the way of the continual assignment of numerical values to these physical properties."

Such 'obvious' assumptions are no longer valid in quantum mechanics. Therefore, other links have to be created between the physical world and the mathematical formalism.

2.2 Hilbert Spaces and Operators

In the following, we briefly review some essential mathematical tools used in quantum mechanics.

Table 2.1. Some relevant properties of vectors and operators in Euclidean and Hilbert spaces

	Euclidean space	Hilbert space
Vectors	$\boldsymbol{r}$	Ψ
Superposition	$\boldsymbol{r} = c_a \boldsymbol{r}_a + c_b \boldsymbol{r}_b$	$\Psi = c_a \Psi_a + c_b \Psi_b$
Scalar product	$\langle \boldsymbol{r}_a | \boldsymbol{r}_b \rangle = \boldsymbol{r}_a \cdot \boldsymbol{r}_b = c_{ab}$	$\langle \Psi_a | \Psi_b \rangle = \langle \Psi_b | \Psi_a \rangle^* = c_{ab}$
	c_a, c_b, c_{ab} real	c_a, c_b, c_{ab} complex
Basis set	$\langle \boldsymbol{v}_i | \boldsymbol{v}_j \rangle = \delta_{ij}$	$\langle \varphi_i | \varphi_j \rangle \equiv \langle i | j \rangle = \delta_{ij}$
Dimension ν	3	$2 \leq \nu \leq \infty$
Completeness	$\boldsymbol{r} = \sum_i x_i \boldsymbol{v}_i$	$\Psi = \sum_i c_i \varphi_i$
Projection	$x_i = \langle \boldsymbol{v}_i | \boldsymbol{r} \rangle$	$c_i = \langle \varphi_i | \Psi \rangle$
Scalar product	$\langle \boldsymbol{r}_a | \boldsymbol{r}_b \rangle = \sum_i x_i^{(a)} x_i^{(b)}$	$\langle \Psi_a | \Psi_b \rangle = \sum_i (c_i^{(a)})^* c_i^{(b)}$
Norm	$\langle \boldsymbol{r} | \boldsymbol{r} \rangle^{1/2} = \left(\sum_i x_i^2 \right)^{1/2}$	$\langle \Psi | \Psi \rangle^{1/2} = \left(\sum_i |c_i|^2 \right)^{1/2}$
Operators	$\hat{R}_\eta(\theta) \, \boldsymbol{r}_a = \boldsymbol{r}_b$	$\hat{Q} \Psi_a = \Psi_b$
Commutators	$[\hat{R}_x(\pi/2), \hat{R}_y(\pi/2)] \neq 0$	$[\hat{Q}, \hat{R}]$
Eigenvalues	$\hat{D}_i \boldsymbol{v}_i = \lambda_i \boldsymbol{v}_i$	$\hat{Q} \varphi_i = q_i \varphi_i$

A Hilbert space is a generalization of the Euclidean, three-dimensional space (see Table 2.1). As in ordinary space, the summation $c_a \Psi_a + c_b \Psi_b$ and the scalar product $\langle \Psi_b | \Psi_a \rangle = c_{ab}$ between two vectors are well defined operations.[3] While the constants c_a, c_b, c_{ab} are real numbers in everyday space, it is essential to allow for complex values in quantum mechanics.

The norm of a vector is defined as the square root of the scalar product of a vector with itself. A vector is said to be normalized if its norm equals one. Two vectors are orthogonal if their scalar product vanishes. A vector Ψ is linearly independent of a subset of vectors $\Psi_a, \Psi_b, \ldots \Psi_d$ if it cannot be expressed as a linear combination of them[4] ($\Psi \neq c_a \Psi_a + c_b \Psi_b + \cdots + c_d \Psi_d$).

These last two concepts allow us to define sets of basis vectors φ_i satisfying the requirement of orthonormalization. Moreover, these sets may be complete, in the sense that any vector Ψ may be expressed as a linear combination of them:[5]

$$\langle \varphi_i | \varphi_j \rangle = \delta_{ij} , \tag{2.2}$$

---

[3] Definitions of these fundamental operations is deferred to each realization of Hilbert spaces [equations (3.2), (3.4) and (4.1), (4.2)]. In the present chapter we use only the fact that they exist and that $\langle a | b \rangle = \langle b | a \rangle^*$.

[4] Although the term 'linear combination' usually refers only to finite sums, we extend its meaning to include also an infinity of terms.

[5] The most familiar case of the expansion of a function in terms of an orthonormal basis set is the Fourier expansion in terms of the exponentials $\exp(ikx)$, which constitutes the complete set of eigenfunctions corresponding to the free particle case (see Sect. 4.3).

$$\Psi = \sum_i c_i \varphi_i , \qquad c_i = \langle \varphi_i | \Psi \rangle \equiv \langle i | \Psi \rangle . \tag{2.3}$$

The scalar product $\langle i | \Psi \rangle$ is the projection of Ψ onto φ_i. The scalar product between two vectors Ψ_a, Ψ_b and the square of the norm of the vector Ψ are also given in terms of the amplitudes c_i in Table 2.1.

The number of states in a basis set is the dimension ν of the associated Hilbert space. It has the value 3 in normal space. In this book, we use Hilbert spaces with dimensions ranging from 2 to a denumerable infinity.

In ordinary space, vectors are defined by virtue of their transformation properties under rotation operations $\hat{R}_\eta(\theta)$ (η denoting the axis of rotation and θ the angle). These operations are generally non-commutative, as the reader may easily verify by performing two successive rotations of $\theta = \pi/2$, first around the x-axis and then around the y-axis, and subsequently comparing the result with the one obtained by reversing the order of these rotations (Fig. 2.2). Vectors in Hilbert spaces may also be transformed through the action of operators $\hat{Q}$ upon them. The operators $\hat{Q}$ obey a non-commutative algebra. We define the commutation operation as

$$[\hat{Q}, \hat{R}] \equiv \hat{Q}\hat{R} - \hat{R}\hat{Q} . \tag{2.4}$$

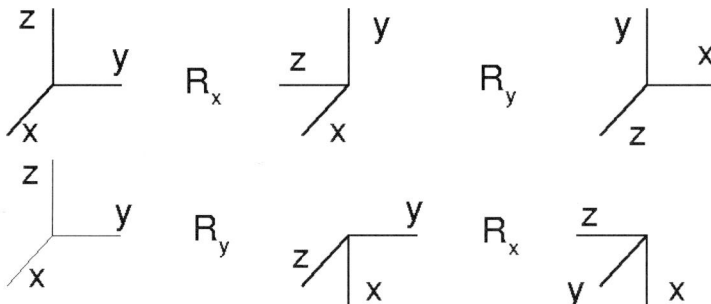

Fig. 2.2. The final orientation of the axes depends on the order of the rotations. R_ν here represents a rotation of $\pi/2$ around the ν-axis

In ordinary space, a dilation $\hat{D}$ is an operation yielding the same vector multiplied by a (real) constant. In Hilbert space, if the state vector $\hat{Q}\varphi_i$ is proportional to φ_i, then φ_i is said to be an eigenvector or an eigenstate of the operator $\hat{Q}$. The constant of proportionality q_i is called the eigenvalue:

$$\hat{Q}\varphi_i = q_i \varphi_i . \tag{2.5}$$

In general, linear combinations of such eigenvectors do not satisfy the eigenvalue equation.

Assume that the operator $\hat{Q}$ acting on the vector Ψ_a generates the vector Φ. The scalar product between another vector Ψ_b and Φ, is called the matrix element of the operator $\hat{Q}$ between the vectors Ψ_a and Ψ_b, and it is symbolically represented as

$$\langle\Psi_b|\Phi\rangle = \langle\Psi_b|\hat{Q}\Psi_a\rangle \equiv \langle\Psi_b|Q|\Psi_a\rangle \equiv \langle b|Q|a\rangle \,. \tag{2.6}$$

The matrix element[6] $\langle\Psi_a|Q|\Psi_a\rangle$ is said to be diagonal: the same vector appears on both sides of the matrix element.

If the operator $\hat{Q}$ belongs to a special class called Hermitian operators, then

$$\langle\varphi_b|Q|\varphi_a\rangle = \langle\varphi_a|Q|\varphi_b\rangle^* \,. \tag{2.7}$$

The eigenvalues q_i of a Hermitian operator are real numbers and the corresponding eigenvectors φ_i constitute a set of basis vectors, as in (2.3). Some more properties of Hermitian operators are listed in Sect. 2.7*.

Up to now we have presented abstract mathematical tools [vectors and (Hermitian) operators], that are used in quantum mechanics. These tools may be represented through concrete, well-known mathematical objects, such as column vectors and matrices (as in Chap. 3), or by means of functions of the coordinate and differential operators (as in Chap. 4).

2.3 Basic Principles of Quantum Mechanics

In this section we present the quantum mechanical relation between the physical world and the mathematical tools that have been outlined in the last section.

It is formulated through the following quantum principles:

Principle 1. *The state of the system[7] is completely described by a vector Ψ – the state vector or state function – belonging to a Hilbert space.*

The state vector Ψ is an abstract entity that carries information about the results of possible measurements. It replaces the classical concepts of position and momentum in the description of physical systems. It is fundamentally different from the electric or magnetic fields in electromagnetic waves, which carry momentum, energy, etc., and in which any externally caused change propagates at a finite, medium-dependent speed.

---

[6] Dirac called the symbols $\langle a|$ and $|a\rangle$ the bra and ket, respectively [16].

[7] This notion of the state of a system is close to the one appearing in classical thermodynamics. It is applied there to many-body systems in which the path of the constituents cannot be traced in practice (for instance, molecules in a gas). However, in quantum mechanics the concept of state is applied even in the case of a single particle.

The state vector may be multiplied by an arbitrary complex constant and still represent the same physical state. Even if we enforce the requirement of normalization, an arbitrary overall phase is left, which has no physical significance. This is not the case for the relative phase of the terms in the sum $c_a \Psi_a + c_b \Psi_b$, which encodes important physical information.

The fact that the sum of two state vectors is another state vector belonging to the same Hilbert space, i.e., describing another state of the system, is usually called the superposition principle. The sum $c_a \Psi_a + c_b \Psi_b$ must not be interpreted in the sense that we have a conglomerate of systems in which some of them are in the state Ψ_a and some in the state Ψ_b, but rather that the system is simultaneously in both component states. This statement is also valid when the system is reduced to a single particle.

This superposition is fundamentally different from any property of classical particles, which are never found as a linear combination of states associated with different trajectories: a tossed coin may fall as head or tails, but not as a superposition of both.

By establishing that the state vector completely describes the state of the system, Principle 1 assumes that there is no way of obtaining information about the system, unless this information is already present in the state vector. However, for a given problem, one may be interested only in some of the degrees of freedom of the system (e.g., magnetic moment, linear momentum, angular momentum, etc.), not in the complete state vector.

The relation between the physical world and states Ψ is more subtle than the classical relation with position and momenta $\boldsymbol{x}, \boldsymbol{p}$. It relies on the following two principles.

Principle 2. *To every physical quantity there corresponds a single operator. In particular, the operators $\hat{x}$ and $\hat{p}$, corresponding to the coordinate and momentum of a particle, satisfy the commutation relation*

$$[\hat{x}, \hat{p}] = i\hbar . \tag{2.8}$$

The commutator is defined in (2.4) and $\hbar$ is Planck's constant h divided by 2π (Table 13.1).[8] The constant $\hbar$ provides an estimate of the domain in which quantum mechanics becomes relevant. It has the dimensions of classical action (energy × time). Classical physics should be applicable to systems in which the action is much larger than $\hbar$.

This is also fundamentally different from classical physics, for which physical properties are identified with numbers (Sect. 2.1).

Since any classical physical quantity may be expressed as a function of coordinate and momentum $Q = Q(x, p)$, the replacement $x \to \hat{x}$ and $p \to \hat{p}$ in the classical expression $Q(x, p)$ yields the operator $\hat{Q} = Q(\hat{x}, \hat{p})$. Thus, a one-to-one correspondence between operators $\hat{Q}$ and physical quantities or

---

[8] This relation has been obtained as a consequence of symmetry transformations in Lorentz space [20].

observables Q is established. However, there are also purely quantum operators, such as the spin operators, that cannot be obtained through such substitution.

The operator corresponding to the classical Hamilton function $H(p,x)$ is called the Hamiltonian. For a conservative system,

$$\hat{H} = \frac{1}{2M}\hat{p}^2 + V(\hat{x}) , \qquad (2.9)$$

where M is the mass of the particle and V the potential.

Principle 3. *The eigenvalues q_i of an operator $\hat{Q}$ constitute the possible results of measurements of the physical quantity Q. The probability[9] of obtaining the eigenvalue q_i is the modulus squared $|c_i|^2$ of the amplitude of the eigenvector φ_i in the state vector Ψ representing the state of the system.*

Since the results of measurements are real numbers, the operators are restricted to be Hermitian (2.34). In particular, the possible values of the energy E_i are obtained by solving the eigenvalue equation

$$\hat{H}\varphi_i = E_i\varphi_i . \qquad (2.10)$$

As in the case of classical mechanics, quantum mechanics may be applied to very different systems, from single-particle and many-body systems to fields. Thus quantum mechanics constitutes a framework in which to develop physical theories, rather than a physical theory by itself. A number of simple, typical, well known problems of a particle moving in a one-dimensional space are discussed in Chaps. 3 and 4.

It is almost as useful to state what the principles do not mean, as to say what they do mean. In the following we quote some common misconceptions regarding quantum states [21].

- "Energy eigenstates are the only allowed ones." This misconception probably arises from the generalized emphasis on the solution of the eigenvalue equation (2.10) and from its similarity to the correct statement: "Energy eigenvalues are the only allowed energies."
- "A state vector describes an ensemble of classical systems." In the standard Copenhagen interpretation, the state vector describes a single system. In none of the acceptable statistical interpretations is the ensemble classical.
- "A state vector describes a single system averaged over some amount of time." The state vector describes a single system at a single instant.

The above three principles are sufficient for the treatment of static situations involving a single particle. Two more principles, concerning many-body systems and the time-evolution of states, are presented in Chaps. 7 and 9, respectively.

---
[9] Notions of probability theory are given in Sect. 2.8*.

2.4 Some Consequences of the Basic Principles

This section displays some consequences of quantum principles in the form of thought experiments. Alternatively, one may obtain the quantum principles as a generalization of the results of such thought experiments (see [16–18]).

Let us consider a Hilbert space consisting of only two independent states $\varphi_{\pm}$. We also assume that these states are eigenstates of an operator $\hat{S}$ corresponding to the eigenvalues ± 1, respectively. Thus the eigenvalue equation $\hat{S}\varphi_{\pm} = \pm \varphi_{\pm}$ is satisfied. The scalar products $\langle \varphi_+ | \varphi_+ \rangle = \langle \varphi_- | \varphi_- \rangle = 1$ and $\langle \varphi_+ | \varphi_- \rangle = 0$ are verified. There are many examples of physical obsevables that may be represented by such operator. For instance, the z-component of the spin (an intrinsic angular momentum carried by the electron and other particles, measured in units of $\hbar/2$)[10] will be frequently used in this book (Sects. 3.1.4, 5.2, 9.2, etc.).

We start by constructing a filter, i.e., an apparatus such that the exiting particles are in a definite eigenstate. In the first part of the apparatus, a beam of particles is split into the two separate $\varphi_{\pm}$ beams, as in the experiment of Otto Stern and Walther Gerlach (Sect. 5.2.1). In the second part, each beam is pushed towards the original direction. Each separate beam may be masked off at the half-way point. Such an apparatus is sketched in Fig. 2.3a, with the φ_- beam masked off. It will be called a φ-filter. It is enclosed within a box drawn with continuous lines.

Any experiment requires first the preparation of the system in some definite initial state. Particles leave the oven in unknown linear combinations Ψ of $\varphi_{\pm}$ states

$$\Psi = \langle \varphi_+ | \Psi \rangle \varphi_+ + \langle \varphi_- | \Psi \rangle \varphi_- \,. \tag{2.11}$$

They are collimated and move along the y-axis. In the following cases, we prepare the particles in the filtered state φ_+, by preventing particles in the state φ_- from leaving the first filter (Fig. 2.3b). In the last stage of the experimental setup we insert another filter as part of the detector, in order to measure the degree of filtration. The detector includes also a photographic plate which records the arriving particles and is observed by an experimentalist (Fig. 2.3c).

In the first experiment, we place the detector immediately after the first filter (Fig. 2.3d). If the φ_- channel is also blocked in the detector, every particle goes through; if the channel φ_+ is blocked, nothing passes. The amplitudes for these processes are $\langle \varphi_+ | \varphi_+ \rangle = 1$ and $\langle \varphi_- | \varphi_+ \rangle = 0$, respectively. The corresponding probabilities, $|\langle \varphi_+ | \varphi_+ \rangle|^2$ and $|\langle \varphi_- | \varphi_+ \rangle|^2$, also are 1 and 0.

We now consider the linear combinations

$$\begin{aligned} \chi_+ &= \langle \varphi_+ | \chi_+ \rangle \varphi_+ + \langle \varphi_- | \chi_+ \rangle \varphi_- \\ \chi_- &= \langle \varphi_+ | \chi_- \rangle \varphi_+ + \langle \varphi_- | \chi_- \rangle \varphi_- \,, \end{aligned} \tag{2.12}$$

---

[10] Another example is given by the polarization states of the photon. Photon polarization, including the experimental setup, is discussed in [22]. Most of the two-state experiments are realized by means of such optical devices.

2.4 Some Consequences of the Basic Principles

where $\langle\chi_+|\chi_+\rangle = \langle\chi_-|\chi_-\rangle = 1$, $\langle\chi_+|\chi_-\rangle = 0$. It is easy to verify that an operator $\hat{R}$, satisfying the eigenvalue equation $\hat{R}\chi_\pm = \pm\chi_\pm$, does not commute with $\hat{S}$. Let us perform the necessary modifications of the second filter so that it can block the particles in either of the states $\chi_\pm$. If $\hat{R}$ corresponds the spin component in the x-direction, the required modification of the detector amounts to a rotation of its filter by an angle $\pi/2$ around the y-axis. Dashed boxes represent filters such that particles exit in the $\chi_\pm$ states (χ-type filters) (Fig. 2.3e).

A particle exiting the first filter in the state φ_+ reorients itself, by chance, within the second filter. The inverse transformation to (2.12) yields

$$\varphi_+ = \langle\varphi_+|\chi_+\rangle^*\chi_+ + \langle\varphi_+|\chi_-\rangle^*\chi_-\,. \tag{2.13}$$

The amplitudes $\langle\chi_+|\varphi_+\rangle = \langle\varphi_+|\chi_+\rangle^*$ and $\langle\chi_-|\varphi_+\rangle = \langle\varphi_+|\chi_-\rangle^*$ are obtained from (2.12). If the χ_- channel of the second filter is blocked, the particle is either projected into the state χ_+ with probability $|\langle\varphi_+|\chi_+\rangle|^2$ or is absorbed with probability $1 - |\langle\varphi_+|\chi_+\rangle|^2 = |\langle\varphi_+|\chi_-\rangle|^2$. This result sounds classical: it is the quantum version of the classical Malus law. However, the projection process is probabilistic. Any information about the previous orientation φ_+ is lost.

We now perform two other experiments which yield results that are spectacularly different from classical expectations. Let us restore the detector filter to the φ-type and introduce a filter of the χ-type between the first filter and the detector (Fig. 2.3f). Thus particles prepared in the φ_+ state exit the second filter in the χ_+ state. In the spin example, particles leave the first filter with the spin pointing in the direction of the positive z-axis, and the second filter pointing along the positive x-axis. The detector measures the number of particles exiting in one of the $\varphi_\pm$ states (spin pointing up or down in the z-direction). The corresponding amplitudes are:

$$\langle\varphi_+|\chi_+\rangle\langle\chi_+|\varphi_+\rangle \tag{2.14}$$
$$\langle\varphi_-|\chi_+\rangle\langle\chi_+|\varphi_+\rangle\,. \tag{2.15}$$

Both components $\varphi_\pm$ may emerge from the detector filter, in spite of the fact that the fraction of the beam in the φ_- state was annihilated inside the first filter. There is no way in classical physics to explain the reconstruction of the beam φ_-. This example illustrates the quantum rule concerning the impossibility of determining two observables associated with operators which do not commute: a precise determination of R destroys the previous information concerning S.

The result of this experiment is also consistent with Principle 1 in Sect. 2.3, since the state vector χ_+ contains all possible information about the system: its past history is not relevant for what happens to it next. The information is lost in the collision with the blocking mask that has been put inside the second filter.

If we repeat the last experiment, but remove the mask from the second filter (Fig. 2.3g), the total amplitude is given by the sum of the amplitudes associated with the two possible intermediate states

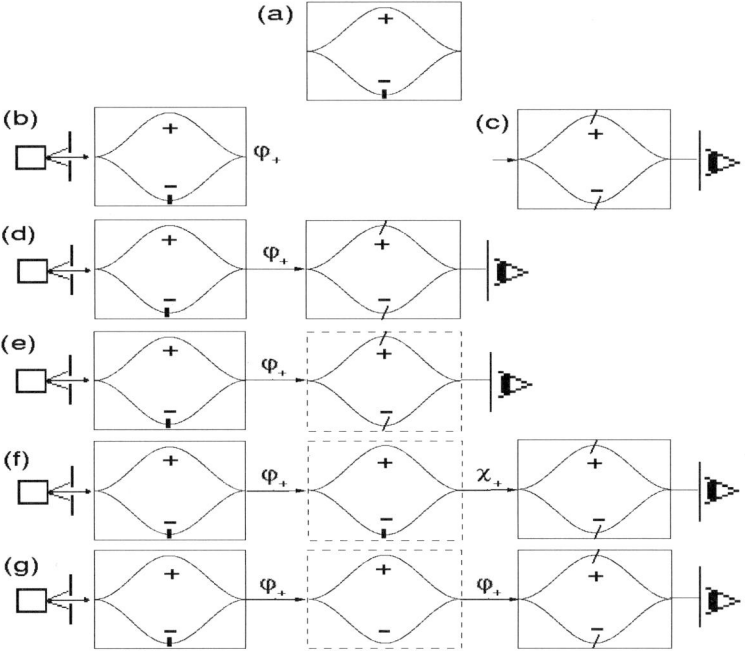

Fig. 2.3. Quantum mechanical thought experiments illustrating the basic principles listed in Sect. 2.3: (a) schematic representation of a filter; (b) preparation of the state of a particle; (c) detector (filter, photoplate, observer); (d)-(g) experiments (see text). The vertical bars denote fixed path blockings, while the diagonal bars indicate paths that can be either opened or closed. For each experiment we perform a measurement in which the upper channel of the detector is open and the down channel blocked, and another measurement with the opposite features

$$\langle\varphi_+|\chi_+\rangle\langle\chi_+|\varphi_+\rangle + \langle\varphi_+|\chi_-\rangle\langle\chi_-|\varphi_+\rangle = \langle\varphi_+|\varphi_+\rangle = 1 \qquad (2.16)$$
$$\langle\varphi_-|\chi_+\rangle\langle\chi_+|\varphi_+\rangle + \langle\varphi_-|\chi_-\rangle\langle\chi_-|\varphi_+\rangle = \langle\varphi_-|\varphi_+\rangle = 0. \qquad (2.17)$$

where the closure property has been applied (2.40). All the particles get through in the first case; none in the second case. In going from the amplitude (2.15) to (2.17) we get fewer particles, despite the fact that more channels are opened.

It is important that none of the intermediate beams suffers an additional disturbance (for example, the influence of an electric field), which may change the relative phases of the two channels.

The result of this last experiment is equivalent to an interference pattern. Classically, such patterns are associated with waves. However, unlike the case of waves, particles here are always detected as lumps of the same size on a screen placed in front of the exit side of the detector filter. No fractions of a lump are ever detected, as befits the behavior of indivisible particles. Therefore, these experiments display wave-particle duality, which is thus accounted for by Principles 1 to 3.

Feynman has commented this result as follows [17], p. 1-1: "We choose to examine a phenomenon which is impossible, *absolutely* impossible, to explain in any classical way, and which has in it the heart of quantum mechanics[11]. In reality, it contains the only[12] mystery."

2.5 Measurement Process

2.5.1 The Concept of Measurement

In this section we specify some basic concepts involved in the process of measurement.[13]

Two or more systems are in interaction if the presence of one leads to changes in the other, and vice versa. Different initial conditions generally lead to different changes, although this may not always be the case.

A measurement is a process in which a system is put in interaction with a piece of apparatus. The apparatus determines the physical quantity or observable to be measured (length, weight, etc.). There is a change in the apparatus which determines the magnitude of the physical quantity. The magnitude has a value if the change can be represented in numerical form.

There are two important steps in a measurement. The first is the preparation of the system to be measured, i.e., the determination of the initial state. In view of the statistical nature of the results obtained in quantum mechanics, it is necessary to prepare many systems in the same initial state. We assume that this is possible.

The second important step, also crucial in the case of quantum systems, is a (macroscopic) change in the apparatus that should be perceptible by a biological cognitive system. In many cases this change is produced by a detector at one end of the apparatus.

For some systems the interaction with the apparatus does not produce a change in the system, or it produces a change which is completely determined. This is the case of classical systems (see Sect. 2.1). On the other hand, this can only be achieved exceptionally in quantum measurements, where a change in the system is generally associated with a change in the apparatus (see Sect. 2.5.2).

---

[11] Matter wave phenomena were experimentally verified for the first time in [5]. Within a continuous effort to test the validity of quantum mechanics, Young's double slit experiments have been performed with electrons, neutrons, atoms, molecules and clusters, including fullerenes (a composite molecule of 64 carbon atoms) [23].

[12] Other fundamental issue in quantum physics is *entanglement* (Sect. 10.1).

[13] Many definitions included here are extracted from [24].

2.5.2 Quantum Measurements

Assume that a measurement of the physical quantity Q, performed on a system in the state Ψ expanded as in (2.3), yields the result q_j. If the same measurement is repeated immediately afterwards, the same value q_j should be obtained with certainty. Thus, the measurement has changed the previous value of the coefficients $c_i \to c_i = \delta_{ij}$. In other words, as a result of the measurement, the system jumps to an eigenstate of the physical quantity that is being measured (the reduction of the state vector). The only exceptions occur when the initial state is already represented by one of the eigenvectors.

Given an initial state vector Ψ, we do not know in general to which eigenstate the system will jump. Only the probabilities, represented by $|c_i|^2$, are determined. This identification of the probabilities is consistent with the following facts:

- their value is always positive,
- their sum is 1 (if the state Ψ is normalized),
- the orthogonality requirement (2.2) ensures that the probability of obtaining any eigenvalue $q_j \neq q_i$ vanishes if the system is initially represented by an eigenstate φ_i (see Table 2.1).

The fact that, given a state vector Ψ, we can only predict the probability $|c_i|^2$ of obtaining eigenvalues q_i constitutes an indeterminacy inherent in quantum mechanics. Our knowledge about the system cannot be improved, for instance, through a second measurement, since the state Ψ has been transformed into φ_i.

The concept of probability implies that we must consider a large number of measurements performed on identical systems, all of them prepared in the same initial state Ψ.

So far we have accepted the reduction interpretation of the measurement process without further discussion. Historically, this was the path followed by most physicists. However, we present one more discussion of the measurement problem in Sect. 11.1.

The diagonal matrix element (2.6) is also called the expectation value or mean value of the operator $\hat{Q}$. It is given by the sum of the eigenvalues weighted by the probability of obtaining them:

$$\langle \Psi | Q | \Psi \rangle = \sum_i q_i |c_i|^2 \,. \tag{2.18}$$

The mean value does not need to be the result q_i of any single measurement, but it is the average value of all the results obtained through the measurement of identical systems.

The uncertainty or standard deviation ΔQ in a given measurement is defined as the square root of the average of the quadratic deviation:

$$\Delta Q = \langle \Psi | \left(Q - \langle \Psi | Q | \Psi \rangle \right)^2 | \Psi \rangle^{1/2}$$
$$= \left(\langle \Psi | Q^2 | \Psi \rangle - \langle \Psi | Q | \Psi \rangle^2 \right)^{1/2} , \qquad (2.19)$$

where

$$\langle \Psi | Q^2 | \Psi \rangle = \sum_i q_i^2 |c_i|^2 . \qquad (2.20)$$

We have postulated the existence of new links between the physical world and mathematics: physical quantities are related to (non-commuting) operators; the state vectors are constructed through operations with these mathematical entities; the feedback to the physical world is made by measurements that yield, as possible results, the eigenvalues of the corresponding operators. An example of this two-way connection between formalism and the physical world is the following. Assume that the system is constructed in a certain physical state, to which the state vector Ψ is assigned. This assignment is tested by means of various probes, i.e., measurements of observables Q, for which we may know the eigenvector φ_i and therefore the probability $|\langle i | \Psi \rangle|^2$ of obtaining the eigenvalues q_i.

This two-way relation between physical world and formalism is not an easy relation [25]: "The most difficult part of learning quantum mechanics is to get a good feeling for how the abstract formalism can be applied to actual phenomena in the laboratory. Such applications almost invariably involve formulating oversimplified abstract models of the real phenomena, to which the quantum formalism can effectively be applied. The best physicists have an extraordinary intuition for what features of the actual phenomena are essential and must be represented in the abstract model, and what features are inessential and can be ignored. It takes years to develop such intuition."

2.6 Commutation Relations and the Uncertainty Principle

In this section it is shown that the commutation relation between two Hermitian operators $\hat{r}, \hat{s}$ determines the precision with which the values of the corresponding physical quantities may be simultaneously determined. Thus Heisenberg uncertainty relations between momenta and coordinates become extended to any pair of observables, and appear as a consequence of their commutation relations.

One assumes two Hermitian operators, $\hat{R}, \hat{S}$, and defines a third (non-Hermitian) operator $\hat{Q}$ such that

$$\hat{Q} \equiv \hat{R} + i\lambda \hat{S} , \qquad (2.21)$$

where λ is a real constant. The minimization with respect to λ of the positively defined norm [see (2.33)]

18   2 Principles of Quantum Mechanics

$$\begin{aligned}0 \leq \langle\hat{Q}\Psi|\hat{Q}\Psi\rangle &= \langle\Psi|Q^+Q|\Psi\rangle \\ &= \langle\Psi|R^2|\Psi\rangle + i\lambda\langle\Psi|[R,S]|\Psi\rangle + \lambda^2\langle\Psi|S^2|\Psi\rangle\,,\end{aligned} \quad (2.22)$$

yields the value

$$\begin{aligned}\lambda_{\min} &= -\frac{i}{2}\langle\Psi|[R,S]|\Psi\rangle/\langle\Psi|S^2|\Psi\rangle \\ &= -\frac{i}{2}\langle\Psi|[R,S]^+|\Psi\rangle^*/\langle\Psi|S^2|\Psi\rangle \\ &= \frac{i}{2}\langle\Psi|[R,S]|\Psi\rangle^*/\langle\Psi|S^2|\Psi\rangle\,.\end{aligned} \quad (2.23)$$

In the second line we have used the definition (2.29) of the Hermitian conjugate. In the last line, the relation $[\hat{R},\hat{S}]^+ = -[\hat{R},\hat{S}]$ stems from the Hermitian character of the operators [see (2.32)]. Substituting the value $\lambda_{\min}$ into (2.22) yields

$$0 \leq \langle\Psi|R^2|\Psi\rangle - \frac{1}{4}\frac{|\langle\Psi|[R,S]|\Psi\rangle|^2}{\langle\Psi|S^2|\Psi\rangle} \quad (2.24)$$

or

$$\langle\Psi|R^2|\Psi\rangle\langle\Psi|S^2|\Psi\rangle \geq \frac{1}{4}|\langle\Psi|[R,S]|\Psi\rangle|^2\,. \quad (2.25)$$

The following two operators $\hat{r},\hat{s}$ have zero expectation value:

$$\hat{r} \equiv \hat{R} - \langle\Psi|R|\Psi\rangle\,, \qquad \hat{s} \equiv \hat{S} - \langle\Psi|S|\Psi\rangle\,, \quad (2.26)$$

and the product of their uncertainties is constrained by [see (2.19)]

$$\Delta r \Delta s \geq \frac{1}{2}|\langle\Psi|[r,s]|\Psi\rangle|\,. \quad (2.27)$$

Operators corresponding to observables can always be written in the form (2.26). If we prepare a large number of quantum systems in the same state Ψ and then perform some measurements of the observable r in some of the systems, and of s in the others, then the standard deviation Δr of the r-results times the standard deviation Δs of the s-results, should satisfy the inequality (2.27).

In the case of coordinate and momentum operators, the relation (2.8) yields the Heisenberg uncertainty relation

$$\Delta x \Delta p \geq \frac{\hbar}{2}\,. \quad (2.28)$$

We emphasize the fact that this relation stems directly from basic principles and, in particular, from the commutation relation (2.8). It constitutes an intrinsic limitation upon our knowledge. This limitation cannot be overcome, for instance, by any improvement of the experiment.

If the state of the system is an eigenstate of the operator $\hat{r}$, then a measurement of the observable r yields the corresponding eigenvalue. The value

2.6 Commutation Relations and the Uncertainty Principle

of the observable s associated with a non-commuting operator $\hat{s}$ is undetermined. This is the case of a plane wave describing a particle in free space (Sect. 4.3) for which the momentum may be determined with complete precision, while the particle is spread over all space.

Another consequence of the relation (2.27) is that the state vector Ψ may be simultaneously an eigenstate of $\hat{r}$ and $\hat{s}$ only if these two operators commute, since in this case the product of their uncertainties vanishes. Moreover, if the operators commute and the eigenvalues of $\hat{s}$ are all different within a subset of states, then the matrix elements of $\hat{r}$ are also diagonal within the same subset of states (see Sect. 2.7*).

Heisenberg conceived the uncertainty relations in order to solve the wave–particle paradox. Pure particle behavior requires localization of the particle, while clear wave behavior appears only when the particle has a definite momentum. Heisenberg's interpretation of this was that each of these extreme classical descriptions is satisfied only when the other is completely untenable. Neither picture is valid for intermediate situations. However, quantum mechanics has to be compatible with the description of the motion of elementary particles (not only with the description of the motion of macroscopic bodies) in terms of trajectories. Heisenberg's answer is that one may construct states Ψ that include a certain amount of localization $\boldsymbol{p}_0(t)$ and $\boldsymbol{x}_0(t)$ in both momentum and coordinate. Thus the motion of a particle has some resemblance to classical motion along trajectories. However, there should be a certain spread in the momentum and in the coordinate, such that the amplitudes $\langle \boldsymbol{p}|\Psi\rangle$ and $\langle \boldsymbol{x}|\Psi\rangle$ in momentum eigenstates and position eigenstates allow the uncertainty relations to hold.

For an illuminating example, Fig. 2.4 displays the capture of a pion by a carbon nucleus [26]. One can determine the mass, energy and charge of the particles, by measuring the length, the grain density, and the scattering

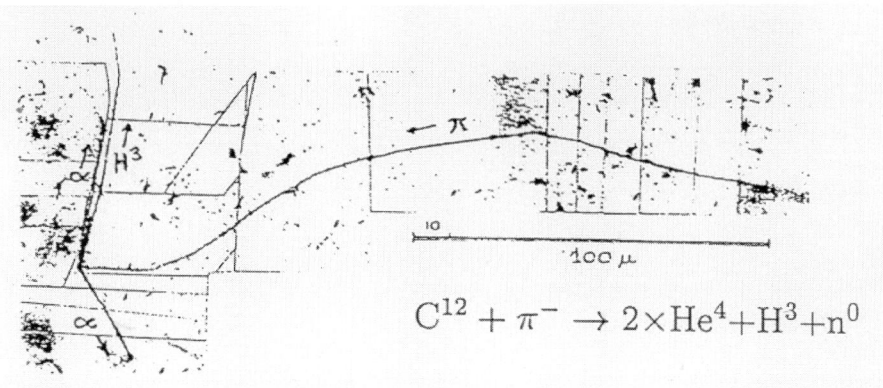

Fig. 2.4. Apparent classical trajectory of a pion (Reproduced with permission from the authors)

direction of their tracks. Let us assume a pion kinetic energy of 10 MeV. Using the pion mass (139 MeV/c^2), one obtains a momentum of $p_\pi = 53$ MeV/c. The uncertainty in the direction perpendicular to the track may be estimated from the width of the track to be ≈ 1 μm, which yields $\Delta p_\perp \approx 10^{-7}$ MeV/c. The ratio $\Delta p_\perp/p_\pi \approx 10^{-9}$ is too small to produce a visible alteration of the apparent trajectory.

2.7* Some Properties of Hermitian Operators

The Hermitian conjugate operator $\hat{Q}^+$ is defined through the equation

$$\langle \Psi_a | Q^+ | \Psi_b \rangle \equiv \langle \Psi_b | Q | \Psi_a \rangle^* , \tag{2.29}$$

referring to the definition (2.6) of the matrix element. Similarly, we may write

$$\langle \Psi_b | Q | \Psi_a \rangle = \langle \hat{Q} \Psi_a | \Psi_b \rangle^* = \langle \hat{Q}^+ \Psi_b | \Psi_a \rangle . \tag{2.30}$$

The following properties are easy to demonstrate:

$$\left(\hat{Q} + c\hat{R} \right)^+ = \hat{Q}^+ + c^* \hat{R}^+ , \tag{2.31}$$

$$\left(\hat{Q} \hat{R} \right)^+ = \hat{R}^+ \hat{Q}^+ . \tag{2.32}$$

According to (2.29), the norm of the state $\hat{Q}\Psi$ is obtained from

$$\langle \hat{Q}\Psi | \hat{Q}\Psi \rangle^{1/2} = \langle \Psi | Q^+ Q | \Psi \rangle^{1/2} . \tag{2.33}$$

The norm is a real, positive number.

An operator is said to be Hermitian if it is equal to its own Hermitian conjugate operator:

$$\hat{Q}^+ = \hat{Q} . \tag{2.34}$$

Assume now that the state φ_i is an eigenstate of the Hermitian operator $\hat{Q}$ corresponding to the eigenvalue q_i. In this case,

$$\begin{aligned}\langle i|Q|i\rangle &= q_i \langle i|i \rangle , \qquad \langle i|Q|i\rangle^* = q_i^* \langle i|i \rangle , \\ \langle i|Q|i\rangle &= \langle i|Q|i\rangle^* \rightarrow q_i = q_i^* .\end{aligned} \tag{2.35}$$

Therefore, the eigenvalues of Hermitian operators are real numbers.

Consider now the non-diagonal terms

$$\langle j|Q|i\rangle = q_i \langle j|i \rangle , \qquad \langle i|Q|j\rangle^* = q_j^* \langle i|j \rangle^* = q_j^* \langle j|i \rangle . \tag{2.36}$$

Then,

$$0 = (q_i - q_j)\langle j|i \rangle , \tag{2.37}$$

i.e., two eigenstates belonging to different eigenvalues are orthogonal. They may also be orthonormal, upon multiplication by an appropriate normalization constant, which is determined up to a phase.

The eigenvectors of a Hermitian operator constitute a complete set of states for a given system. This means that any state function Ψ, describing any state of the same system, may be expressed as a linear combination of basis states φ_i [see (2.3)].

We define the projection operator (a theoretical filter) $|i\rangle\langle i|$ through the equation

$$|i\rangle\langle i|\varphi_j \equiv \langle i|j\rangle\varphi_i = \delta_{ij}\varphi_i \,, \tag{2.38}$$

which implies that

$$\sum_i |i\rangle\langle i|\Psi = \Psi \,, \tag{2.39}$$

for any Ψ. Thus unity may be expressed as the operator $\sum_i |i\rangle\langle i|$. From this property stems the closure property, according to which the matrix elements of the product of Hermitian operators may be calculated as the sum over all possible intermediate states of products of the matrix elements corresponding to each separate operator:

$$\langle i|QR|j\rangle = \sum_k \langle i|Q|k\rangle\langle k|R|j\rangle \,. \tag{2.40}$$

2.8* Notions of Probability Theory

Probability theory studies the likelihood P_i that the outcome q_i of an event will take place. The limits of P_i are

$$0 \leq P_i \leq 1 \,. \tag{2.41}$$

If $P_i = 0$, the outcome q_i cannot occur; if $P_i = 1$, it will take place with certainty.

If two events (i, j) are statistically independent, the probability that both i and j take place is given by the product

$$P_{(i \text{ and } j)} = P_i P_j \,. \tag{2.42}$$

If two events are mutually exclusive, the probability that one or the other occur is the sum

$$P_{(i \text{ or } j)} = P_i + P_j \,. \tag{2.43}$$

Probability may be defined as

$$P_i \equiv \lim_{N \to \infty} \frac{n_i}{N} \,, \tag{2.44}$$

where n_i is the number of outcomes q_i of a total of $N \equiv \sum_i n_i$ outcomes. Since the limit $N \to \infty$ is never attained, N should in practice be made large enough to ensure that the fluctuations become sufficiently small.

The collection of P_is is called the (discrete) probability distribution. The concepts of average $\langle Q \rangle$, root mean square $\langle Q^2 \rangle^{1/2}$ and root mean square deviation ΔQ, applied in Sect. 2.5, are given by

$$\langle Q \rangle = \sum_i q_i P_i ,$$

$$\langle Q^2 \rangle^{1/2} = \left(\sum_i q_i^2 P_i \right)^{1/2} , \qquad (2.45)$$

$$\Delta Q = \langle (Q - \langle Q \rangle)^2 \rangle^{1/2} = \left(\langle Q^2 \rangle - \langle Q \rangle^2 \right)^{1/2} .$$

In the case of a continuous distribution, the sums $\sum_i$ are replaced by integrals $\int \mathrm{d}x$. Instead of probabilities P_i, one defines probability densities $\rho(x)$ such that

$$1 = \int_{-\infty}^{\infty} \rho(x) \mathrm{d}x , \qquad \langle Q \rangle = \int_{-\infty}^{\infty} q(x) \rho(x) \mathrm{d}x . \qquad (2.46)$$

Problems

Problem 1. Assume that the state Ψ is given by the linear combination $\Psi = c_1 \Psi_1 + c_2 \Psi_2$, where the amplitudes c_1, c_2 are arbitrary complex numbers, and both states Ψ_1, Ψ_2 are normalized.

1. Normalize the state Ψ, assuming that $\langle 1|2 \rangle = 0$.
2. Find the probability of the system being in the state Ψ_1.

Problem 2. Use the same assumptions as in Problem 1, but $\langle 1|2 \rangle = c \neq 0$.

1. Find a linear combination $\Psi_3 = \lambda_1 \Psi_1 + \lambda_2 \Psi_2$ that is orthogonal to Ψ_1 and normalized.
2. Express the vector Ψ as a linear combination of Ψ_1 and Ψ_3.

Problem 3. Prove the equations (2.31) and (2.32). Hint: apply successively the definition of Hermitian conjugate to the operators $\hat{Q}, \hat{R}$. For instance, start with $\langle \Psi_b | QR | \Psi_a \rangle = \langle \hat{R} \Psi_a | Q^+ | \Psi_b \rangle^*$.

Problem 4. Show that

$$[\hat{Q}, \hat{R}] = -[\hat{R}, \hat{Q}] , \qquad [\hat{Q}\hat{R}, \hat{S}] = [\hat{Q}, \hat{S}]\hat{R} + \hat{Q}[\hat{R}, \hat{S}] . \qquad (2.47)$$

Problem 5. Find the commutation relation between the coordinate operator $\hat{x}$ and the one-particle Hamiltonian (2.9). Discuss the result in terms of the simultaneous determination of energy and position of a particle.

Problem 6. Find the commutation relation $[\hat{p}^n, \hat{x}]$, where n is an integer.

Problem 7. Verify that the commutation relation (2.8) is consistent with the fact that the operators $\hat{x}$ and $\hat{p}$ are Hermitian.

Problem 8. Assume the basis set of states φ_i.
1. Calculate the effect of the operator $\hat{R} \equiv \sum_i |i\rangle\langle i|$ on an arbitrary state Ψ.
2. Repeat for the operator $\hat{R} \equiv \prod_i (\hat{Q} - q_i)$, assuming that the equation $\hat{Q}\varphi_i = q_i \varphi_i$ is satisfied.

Problem 9. Find the relation between the matrix elements of the operators $\hat{p}$ and $\hat{x}$ in the basis of eigenvectors of the Hamiltonian (2.9).

Problem 10. Consider a basis set of states made up of three vector states. The operator $\hat{Q}$ satisfies the equations
$$\hat{Q}\varphi_1 = q\varphi_1, \qquad \hat{Q}\varphi_2 = 2q\varphi_2, \qquad \hat{Q}\varphi_3 = 2q\varphi_3.$$
Assume that the system is prepared in the state
$$\Psi = \frac{1}{\sqrt{6}}(\varphi_1 + \varphi_2) + \sqrt{\frac{2}{3}}\varphi_3.$$

1. Calculate the expectation value of the observable Q.
2. What are the possible results of a measurement of Q, and what are their respective probabilities?
3. What is the vector state after a measurement of Q that has yielded the value $2q$?

Problem 11. Evaluate in m.k.s. units possible values of the precision to which the velocity and the position of a car should be measured in order to verify the uncertainty relation (2.28).

Problem 12. A 10 MeV proton beam is collimated by means of diaphragms with a 5 mm aperture.
1. Show that the spread in energy ΔE_H associated with the uncertainty principle is negligible relative to the total spread $\Delta E \approx 10^{-3}$ MeV.
2. Calculate the distance x that a proton has to travel in order to traverse 5 mm in a perpendicular direction, if the perpendicular momentum is due only to the uncertainty principle.

3 The Heisenberg Realization of Quantum Mechanics

In this chapter we present the simplest realization of the basic principles of quantum mechanics. We employ column vectors as state vectors and square matrices as operators. This formulation is especially suitable for Hilbert spaces with finite dimensions. However, we also treat the problem of the harmonic oscillator within this framework.

3.1 Matrix Formalism

3.1.1 A Realization of the Hilbert Space

The state vector Ψ may be expressed by means of the amplitudes c_i filling the successive rows of a column vector:

$$\Psi = (c_i) \equiv \begin{pmatrix} c_a \\ c_b \\ \vdots \\ c_\nu \end{pmatrix} . \tag{3.1}$$

The dimension of the Hilbert space is given by ν. The sum of two column vectors is another column vector in which the amplitudes are added:

$$\alpha_B \Psi_B + \alpha_C \Psi_C = (\alpha_B b_i + \alpha_C c_i) . \tag{3.2}$$

The scalar product requires the definition of the adjoint vector Ψ^+, i.e., a row vector obtained from Ψ with amplitudes

$$\Psi^+ = (c_a^*, c_b^*, \ldots, c_\nu^*) . \tag{3.3}$$

The scalar product of two vectors Ψ_B and Ψ_C is defined as the product of the adjoint vector Ψ_B^+ and the vector Ψ_C, viz.,

$$\langle \Psi_B | \Psi_C \rangle = \sum_{i=a}^{i=\nu} b_i^* c_i ,$$

$$\langle\Psi|\Psi\rangle = \sum_{i=a}^{i=\nu} |c_i|^2 = 1 \, . \tag{3.4}$$

A useful set of (orthonormal) basis states is given by the vector columns φ_i with amplitudes $c_j = \delta_{ij}$. In such a basis, the arbitrary vector (3.1) may be expanded as

$$\Psi = c_a \begin{pmatrix} 1 \\ 0 \\ \vdots \\ 0 \end{pmatrix} + c_b \begin{pmatrix} 0 \\ 1 \\ \vdots \\ 0 \end{pmatrix} + \cdots + c_\nu \begin{pmatrix} 0 \\ 0 \\ \vdots \\ 1 \end{pmatrix} . \tag{3.5}$$

All the properties listed in Table 2.1 are reproduced within the framework of column vectors.

Operators are represented by square matrices, which are denoted as

$$\hat{Q} \to \mathcal{Q} = (\langle i|Q|j\rangle) \equiv \begin{pmatrix} \langle a|Q|a\rangle & \langle a|Q|b\rangle & \cdots & \langle a|Q|\nu\rangle \\ \langle b|Q|a\rangle & \langle b|Q|b\rangle & \cdots & \langle b|Q|\nu\rangle \\ \vdots & \vdots & \ddots & \vdots \\ \langle \nu|Q|a\rangle & \langle \nu|Q|b\rangle & \cdots & \langle \nu|Q|\nu\rangle \end{pmatrix} . \tag{3.6}$$

The matrix elements $\langle i|Q|j\rangle$ are constructed as in (2.6). The matrices corresponding to physical observables are Hermitian [see (2.7)]. The initial state j labels the columns, while the final state i labels the rows. The order $a, b, \ldots, \nu$ is immaterial, provided it is the same in both columns and rows (i.e., the matrix elements $\langle i|Q|i\rangle$ should lie on the diagonal).

A matrix multiplying a vector yields another vector, so that

$$\Psi_B = \mathcal{Q}\Psi_C \longleftrightarrow b_i = \sum_j \langle i|Q|j\rangle c_j \, . \tag{3.7}$$

The product of two matrices is another matrix:

$$\mathcal{S} = \mathcal{Q}\mathcal{R} \longleftrightarrow \langle i|S|j\rangle = \sum_k \langle i|Q|k\rangle \langle k|R|j\rangle \, , \tag{3.8}$$

which is consistent with the closure property (2.40). The multiplication of matrices is a non-commutative operation, as befits the representation of quantum operators.

3.1.2 The Solution of the Eigenvalue Equation

In matrix form, the eigenvalue equation (2.5) reads

$$\begin{pmatrix} \langle a|Q|a\rangle & \langle a|Q|b\rangle & \cdots & \langle a|Q|\nu\rangle \\ \langle b|Q|a\rangle & \langle b|Q|b\rangle & \cdots & \langle b|Q|\nu\rangle \\ \vdots & \vdots & \ddots & \vdots \\ \langle \nu|Q|a\rangle & \langle \nu|Q|b\rangle & \cdots & \langle \nu|Q|\nu\rangle \end{pmatrix} \begin{pmatrix} c_a \\ c_b \\ \vdots \\ c_\nu \end{pmatrix} = q \begin{pmatrix} c_a \\ c_b \\ \vdots \\ c_\nu \end{pmatrix} , \tag{3.9}$$

which is equivalent to the ν linear equations (one equation for each value of i)

$$\sum_{j=1}^{j=\nu}\langle i|Q|j\rangle c_j = qc_i \ . \tag{3.10}$$

The eigenvalues q and the amplitudes c_i are the unknowns to be determined.[1]

The solution to (3.10) is obtained by casting the original matrix ($\langle i|Q|j\rangle$) into a diagonal form. In this case the diagonal matrix elements become the eigenvalues, $\langle i|Q|j\rangle = \delta_{ij}q_i$. The ith eigenvector is given by the amplitudes $c_j = \delta_{ij}$, as in (3.5). For instance,

$$\begin{pmatrix} q_1 & 0 & \cdots & 0 \\ 0 & q_2 & \cdots & 0 \\ \vdots & \vdots & \ddots & \vdots \\ 0 & 0 & \cdots & q_\nu \end{pmatrix} \begin{pmatrix} 0 \\ 1 \\ \vdots \\ 0 \end{pmatrix} = q_2 \begin{pmatrix} 0 \\ 1 \\ \vdots \\ 0 \end{pmatrix} \ . \tag{3.11}$$

The linear homogeneous equations (3.10) have the trivial solution $c_i = 0$, to be discarded. The existence of additional, non-trivial solutions requires the determinant to vanish:

$$\det\left(\langle i|Q|j\rangle - q\delta_{ij}\right) = 0 \ . \tag{3.12}$$

This eigenvalue equation is equivalent to a polynomial equation for q. Its ν roots are the eigenvalues of the operator $\hat{Q}$.

The vanishing of the determinant (3.12) implies that one of the equations (3.10) may be expressed as a linear combination of the other $\nu - 1$ equations. Therefore, by disregarding one of these equations (for instance, the one corresponding to the νth row) and dividing the remaining equations by c_a, one obtains a set of $\nu - 1$ non-homogeneous linear equations[2] yielding the value of the ratios $c_b/c_a, c_c/c_a, \ldots, c_\nu/c_a$, for each eigenvalue q. The normalization equation (3.4) determines the value of $|c_a|^2$, up to the usual overall arbitrary phase of the state vector. Note that the relative phases in the linear combination have physical significance, although the overall phase is unimportant.

3.1.3 Transformation Between Sets of Basis States

Diagonalization yields a new set of eigenstates ϕ_a. Each of them may be expressed as a linear combination of the old basis states φ_i

---

[1] This equation may be obtained directly by using the expansion (2.3) on both sides of the general eigenvalue equation $\hat{Q}\Psi = q\Psi$. One obtains $\sum_j c_j \hat{Q}\varphi_j = q\sum_j c_j \varphi_j$. The scalar product with φ_i of both sides of this last equation yields (3.10).

[2] If several roots have the same eigenvalue, more equations should be discarded in order to get a non-homogeneous set of equations.

$$\phi_a = \sum_i \langle i|a\rangle \varphi_i . \tag{3.13}$$

The amplitudes $\langle i|a\rangle$ are the matrix elements of a matrix $\mathcal{U} = (\langle i|a\rangle)$ that transforms the basis set φ_i into the basis set ϕ_a. Such a matrix does not represent a physical observable and it is not therefore required to be Hermitian.

The inverse transformation to (3.13) is written as

$$\varphi_i = \sum_a \langle a|i\rangle \phi_a . \tag{3.14}$$

Therefore, the inverse transformation $\mathcal{U}^{-1}$ is the transposed conjugate:

$$\mathcal{U}^{-1} = (\langle a|i\rangle) = \mathcal{U}^+ , \qquad \mathcal{U}^+\mathcal{U} = \mathcal{U}\mathcal{U}^+ = \mathcal{I} , \tag{3.15}$$

where $\mathcal{I}$ is the unit matrix. Equation (3.15) implies that

$$\sum_i \langle a|i\rangle \langle i|b\rangle = \langle a|b\rangle = \delta_{ab} ,$$
$$\sum_a \langle i|a\rangle \langle a|j\rangle = \langle i|j\rangle = \delta_{ij} . \tag{3.16}$$

A matrix satisfying (3.15) is said to be unitary. If states are transformed according to $\phi = \mathcal{U}\varphi$, then the state $\mathcal{U}\hat{Q}\varphi$ ($\hat{Q}$ = physical operator) may be written as

$$\mathcal{U}\hat{Q}\varphi = \mathcal{U}\hat{Q}\mathcal{U}^+\mathcal{U}\varphi = \hat{R}\phi , \tag{3.17}$$

which yields the rule for the transformation of operators, namely

$$\hat{R} = \mathcal{U}\hat{Q}\mathcal{U}^+ . \tag{3.18}$$

Unitary transformations preserve the value of the determinant and the trace:

$$|(\langle a|R|b\rangle)| = |(\langle i|Q|j\rangle)| ,$$
$$\text{trace}\,(Q) \equiv \sum_i \langle i|Q|i\rangle = \text{trace}\,(R) \equiv \sum_a \langle a|R|a\rangle . \tag{3.19}$$

The modulus squared $|\langle a|i\rangle|^2$ is both the probability of measuring the eigenvalue q_i, associated with the eigenstate φ_i, if the system is in the state ϕ_a, and the probability of measuring the eigenvalue r_a, associated with the eigenstate ϕ_a, when the state of the system is φ_i.

3.1.4 Application to 2×2 Matrices

An example of a diagonal matrix is the matrix representing the z-component of the spin operator (Sect. 5.2.2):

3.1 Matrix Formalism

$$\hat{S}_z = \frac{\hbar}{2}\begin{pmatrix} 1 & 0 \\ 0 & -1 \end{pmatrix}. \tag{3.20}$$

The eigenvectors corresponding to spin up and spin down are, respectively,

$$\varphi_{\uparrow z} \equiv \varphi_{(s_z=\hbar/2)} = \begin{pmatrix} 1 \\ 0 \end{pmatrix}, \quad \varphi_{\downarrow z} \equiv \varphi_{(s_z=-\hbar/2)} = \begin{pmatrix} 0 \\ 1 \end{pmatrix}. \tag{3.21}$$

Let us diagonalize a general Hermitian matrix of order two:

$$\begin{pmatrix} \langle a|Q|a\rangle & \langle a|Q|b\rangle \\ \langle b|Q|a\rangle & \langle b|Q|b\rangle \end{pmatrix}. \tag{3.22}$$

The resulting eigenvalues are

$$q_\pm = \frac{1}{2}\left(\langle a|Q|a\rangle + \langle b|Q|b\rangle\right) \pm \frac{1}{2}\sqrt{(\langle a|Q|a\rangle - \langle b|Q|b\rangle)^2 + 4|\langle a|Q|b\rangle|^2}, \tag{3.23}$$

while the amplitudes of the eigenvectors are given by

$$\left.\frac{c_b}{c_a}\right|_\pm = \frac{q_\pm - \langle a|Q|a\rangle}{\langle a|Q|b\rangle}, \quad (c_a)_\pm = \left(1 + \left|\frac{c_b}{c_a}\right|_\pm^2\right)^{-\frac{1}{2}}. \tag{3.24}$$

Figure 3.1 plots the eigenvalues $q_\pm$ and the initial expectation values as a function of $Q \equiv \langle a|Q|a\rangle$, assuming a traceless situation ($\langle a|Q|a\rangle = -\langle b|Q|b\rangle$). The eigenvalue q_+ is always higher than $|Q|$, while q_- is always below $-|Q|$: the two eigenvalues repel each other and never cross, if $\langle a|Q|b\rangle \neq 0$. The distance $\Delta = \sqrt{Q^2 + |\langle a|Q|b\rangle|^2} - Q$ measures the increase in the eigenvalue of Q, due to the superposition of the states φ_a, φ_b, and it is maximized at the crossing point $Q = 0$.

The physical world displays many systems with two states.[3] An example of this is an electron and two protons. As a reasonable approximation, we may neglect the motion of the protons, since they are much heavier than the electron. The two states φ_a, φ_b represent the electron bound to each of the protons: a hydrogen atom and a separate proton in each case. In this case the Hamiltonian $\hat{H}$ plays the role of $\hat{Q}$ in (3.22) and (3.23). The extra binding Δ, arising from the superposition of states φ_a, φ_b, allows for the existence of a bound state: the stability of the ionized hydrogen molecule thus has a purely quantum mechanical origin. This problem is discussed in more detail in Sect. 8.4.1.

The calculations (3.23) and (3.24) are quickly made using the matrix associated with the spin component $\hat{S}_x$ (5.23):

$$\hat{S}_x = \frac{\hbar}{2}\begin{pmatrix} 0 & 1 \\ 1 & 0 \end{pmatrix}. \tag{3.25}$$

---

[3] In fact, any two states sufficiently isolated from the remaining ones may be approximated as a two-state system, for which the no-crossing rule holds. See also Sect. 2.4

30 3 The Heisenberg Realization of Quantum Mechanics

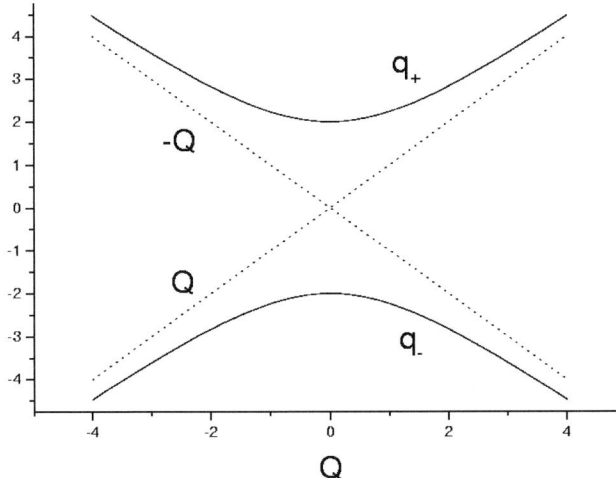

Fig. 3.1. Eigenvalues $q_\pm$ of a 2×2 system (*continuous curves*) as a function of Q, half of the energy distance between the diagonal matrix elements (*dotted lines*)

Equation (3.23) yields the eigenvalues $s_x = \pm\hbar/2$ and the eigenvectors

$$\phi_{\uparrow x} = \frac{1}{\sqrt{2}}\begin{pmatrix}1\\1\end{pmatrix} = \frac{1}{\sqrt{2}}\varphi_{\uparrow z} + \frac{1}{\sqrt{2}}\varphi_{\downarrow z},$$

$$\phi_{\downarrow x} = \frac{1}{\sqrt{2}}\begin{pmatrix}1\\-1\end{pmatrix} = \frac{1}{\sqrt{2}}\varphi_{\uparrow z} - \frac{1}{\sqrt{2}}\varphi_{\downarrow z}. \quad (3.26)$$

These equations express the eigenstates of $\hat{S}_x$ as linear combinations of the eigenstates (3.21) of $\hat{S}_z$.

The unitary transformation

$$\mathcal{U} = \frac{1}{\sqrt{2}}\begin{pmatrix}1 & 1\\1 & -1\end{pmatrix} \quad (3.27)$$

transforms the basis set of eigenvectors of the operator $\hat{S}_z$ into the basis set of eigenvectors of $\hat{S}_x$, in accordance with (3.13):

$$\mathcal{U}\begin{pmatrix}1\\0\end{pmatrix} = \frac{1}{\sqrt{2}}\begin{pmatrix}1\\1\end{pmatrix}, \quad \mathcal{U}\begin{pmatrix}0\\1\end{pmatrix} = \frac{1}{\sqrt{2}}\begin{pmatrix}1\\-1\end{pmatrix}. \quad (3.28)$$

Similarly, the operator $\hat{S}_z$ is transformed into the operator $\hat{S}_x$ [see (3.18)]:

$$\begin{pmatrix}0 & 1\\1 & 0\end{pmatrix} = \mathcal{U}\begin{pmatrix}1 & 0\\0 & -1\end{pmatrix}\mathcal{U}^+. \quad (3.29)$$

The trace of the matrix representing the spin operator S_x is invariant (and equal to 0) under the unitary transformation. In this spin case, it happens that the eigenvalues of the operators $\hat{S}_z$ and $\hat{S}_x$ are the same. This result is to be expected for physical reasons: the eigenvalues do indeed have physical significance, while the orientation of the coordinate system in an isotropic space does not.

3.2 Harmonic Oscillator

Here we present a solution to the harmonic oscillator problem, a solution that stems directly from the basic principles listed in Sect. 2.3. The Hamiltonian corresponding to the one-dimensional harmonic oscillator is

$$\hat{H} = \frac{1}{2M}\hat{p}^2 + \frac{M\omega^2}{2}\hat{x}^2 , \qquad (3.30)$$

where ω is the classical frequency (Fig. 3.2).

The harmonic oscillator potential is probably the most widely used potential in physics, because of its ability to represent physical potentials in the vicinity of stable equilibrium [e.g., vibrational motion in molecules (8.30)].

It is always convenient to start by finding the order of magnitude of the quantities involved. To do so, we apply the Heisenberg uncertainty principle (2.28). If the substitutions $x \to \Delta_x$ and $\hat{p} \to \hbar/2\Delta_x$ are made in the harmonic oscillator energy, so that

$$E \geq \frac{\hbar^2}{8M\Delta_x^2} + \frac{M\omega^2\Delta_x^2}{2} , \qquad (3.31)$$

then minimization with respect to Δ_x gives the value at the minimum:

$$(\Delta_x)_{\min} = \sqrt{\frac{\hbar}{2M\omega}} , \qquad (3.32)$$

which yields the characteristic orders of magnitude

$$x_c = \sqrt{\frac{\hbar}{M\omega}} , \qquad p_c = \sqrt{\hbar M\omega} , \qquad E_c = \hbar\omega . \qquad (3.33)$$

The classical equilibrium position $x = p = 0$ is not compatible with the uncertainty principle, because it implies a simultaneous determination of coordinate and momentum. The replacement of Δx in (3.31) with (3.32) yields the zero-point energy,[4] i.e., the minimum energy that the harmonic oscillator may have, namely,

$$E_0 = \frac{1}{2}\hbar\omega . \qquad (3.34)$$

---

[4] The procedure is only expected to yield correct orders of magnitude. It is a peculiarity of the harmonic oscillator that the results are exact.

3.2.1 Solution of the Eigenvalue Equation

We intend to solve (2.10). The unknowns are the eigenvalues E_i and the eigenfunctions φ_i. The fundamental tool entering the present solution is the commutation relation (2.8).

We first define the operators a^+, a

$$a^+ \equiv \sqrt{\frac{M\omega}{2\hbar}}\hat{x} - \frac{i}{\sqrt{2M\hbar\omega}}\hat{p}, \qquad a \equiv \sqrt{\frac{M\omega}{2\hbar}}\hat{x} + \frac{i}{\sqrt{2M\hbar\omega}}\hat{p}. \tag{3.35}$$

The operators $\hat{x}$ and $\hat{p}$ are Hermitian, since they correspond to physical observables. Therefore the operators a, a^+ are Hermitian conjugates of each other, according to (2.31). They satisfy the commutation relations

$$\left[\hat{H}, a^+\right] = \hbar\omega a^+, \tag{3.36}$$

$$\left[a, a^+\right] = 1. \tag{3.37}$$

We now construct the matrix elements (2.6) for both sides of (3.36), making use of two eigenstates φ_i, φ_j:

$$\langle i|[H, a^+]|j\rangle = (E_i - E_j)\langle i|a^+|j\rangle = \hbar\omega\langle i|a^+|j\rangle. \tag{3.38}$$

We conclude that the matrix element $\langle i|a^+|j\rangle$ vanishes, unless the difference $E_i - E_j$ between the energies of the two eigenstates is the constant $\hbar\omega$. This fact implies that we may sequentially order the eigenstates connected by a^+, the difference between two consecutive energies being $\hbar\omega$. Another consequence is that we may assign an integer number n to each eigenstate.

Since a, a^+ are Hermitian conjugate operators, we may also write

$$\langle n+1|a^+|n\rangle = \langle n|a|n+1\rangle^*. \tag{3.39}$$

Finally, we expand the expectation value of (3.37):

$$\begin{aligned}
1 &= \langle n|[a, a^+]|n\rangle \\
&= \langle n|a|n+1\rangle\langle n+1|a^+|n\rangle - \langle n|a^+|n-1\rangle\langle n-1|a|n\rangle \\
&= |\langle n+1|a^+|n\rangle|^2 - |\langle n|a^+|n-1\rangle|^2.
\end{aligned} \tag{3.40}$$

This is a finite difference equation in $y_n = |\langle n+1|a^+|n\rangle|^2$, of the type $1 = y_n - y_{n-1}$. Its solutions are

$$|\langle n+1|a^+|n\rangle|^2 = n + c, \tag{3.41}$$

where c is a constant. Since the left-hand side is positive definite, the quantum number n must have a lower limit, which we may choose to be $n = 0$. It corresponds to the ground state φ_0. In such a case, the matrix element $\langle 0|a^+|-1\rangle$ should disappear, which fixes the value of the constant $c = 1$. Therefore, according to (3.39), $\langle -1|a|0\rangle = 0$, which is equivalent to

$$a\varphi_0 = 0 , \tag{3.42}$$

i.e., the ground state is annihilated by the operator a, which is called the annihilation operator.

The whole set of orthogonal eigenstates may be constructed by repeatedly applying the operator a^+, the creation operator.

$$\varphi_n = \frac{1}{\sqrt{n!}} \left(a^+\right)^n \varphi_0 , \qquad n = 0, 1, \ldots . \tag{3.43}$$

These states are labeled with the quantum number n. They are eigenstates of the operator $\hat{n} = a^+ a$, the number operator, with eigenvalues n:

$$\hat{n}\varphi_n = \frac{1}{\sqrt{n!}} a^+ [a, (a^+)^n]\varphi_0 = \frac{1}{\sqrt{n!}} a^+ n \left(a^+\right)^{n-1} \varphi_0 = n\varphi_n . \tag{3.44}$$

The factor $1/\sqrt{n!}$ ensures the normalization of the eigenstates.

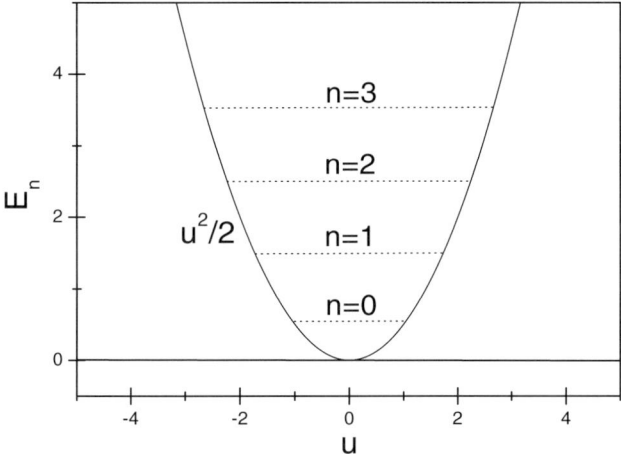

Fig. 3.2. Harmonic oscillator potential and its eigenvalues. All energies are given in units of $\hbar\omega$. The dimensionless variable $u = x/x_c$ has been used

In order to find the matrix elements of the operators $\hat{x}$ and $\hat{p}$, we invert the definition in (3.35):

$$\hat{x} = \sqrt{\frac{\hbar}{2M\omega}} \left(a^+ + a\right) , \qquad \hat{p} = \mathrm{i}\sqrt{\frac{M\hbar\omega}{2}} \left(a^+ - a\right) , \tag{3.45}$$

and obtain the non-vanishing matrix elements

$$\langle n+1|x|n\rangle = \langle n|x|n+1\rangle = \sqrt{\frac{\hbar}{M\omega} \frac{n+1}{2}} , \tag{3.46}$$

$$\langle n+1|p|n\rangle = \langle n|p|n+1\rangle^* = \mathrm{i}\sqrt{M\hbar\omega \frac{n+1}{2}} . \tag{3.47}$$

Substitution of (3.45) into the Hamiltonian yields

$$\hat{H} = \hbar\omega \left(\hat{n} + \frac{1}{2}\right) . \tag{3.48}$$

The Hamiltonian matrix is thus diagonal, with eigenvalues E_n represented in Fig. 3.2

$$\langle n|H|n\rangle = E_n = \hbar\omega \left(n + \frac{1}{2}\right) . \tag{3.49}$$

The creation and annihilation operators are often used in many-body quantum physics (Sect. 7.5*). They are also essential tools in quantum field theory, since they allow us to represent the creation and annihilation of phonons, photons, mesons, etc. (Sects. 9.5.2 and 9.5.3).

Quantum mechanics has provided the present derivation through the fundamental commutation relation (2.8), yielding the properties of the matrix elements $\langle n|a^+|m\rangle$ in a straightforward way. The results are also valid for any problem involving two operators satisfying (2.8), with a Hamiltonian that is quadratic in these operators.

3.2.2 Some Properties of the Solution

In the following we use this exact, analytical solution of the harmonic oscillator problem to deduce some relevant features of quantum mechanics.[5] The discussion of the spatial dependence of the harmonic oscillator problem is deferred to Sect. 4.2.

By using the closure property (2.40) and the matrix elements (3.46) and (3.47), one obtains the matrix element of the commutator $[\hat{x}, \hat{p}]$:

$$\begin{aligned}\langle n|[x,p]|m\rangle &= \langle n|x|n+1\rangle\langle n+1|p|m\rangle + \langle n|x|n-1\rangle\langle n-1|p|m\rangle \\ &\quad - \langle n|p|n+1\rangle\langle n+1|x|m\rangle - \langle n|p|n-1\rangle\langle n-1|x|m\rangle \\ &= i\hbar\delta_{nm} .\end{aligned} \tag{3.50}$$

The matrix elements of the operators $\hat{x}^2$ and $\hat{p}^2$ may be constructed in a similar way:

$$\frac{M\omega}{\hbar}\langle n|x^2|n\rangle = \frac{1}{\hbar M\omega}\langle n|p^2|n\rangle = n + \frac{1}{2} , \tag{3.51}$$

which implies the equality between the kinetic energy and potential expectation values (virial theorem).

Applying the definition of the root mean square deviation ΔQ given in (2.19), the product $\Delta x \Delta p$ yields

---

[5] However, the reader is warned against concluding that most quantum problems are analytically solvable, a conclusion that may be reinforced throughout these notes by the repeated utilization of exactly soluble examples. Most quantum problems require insight into physics to approximate the solution and/or sizeable computing facilities.

$$(\Delta x)_n (\Delta p)_n = \frac{E_n}{\omega} = \hbar \left(n + \frac{1}{2}\right) \geq \frac{1}{2}\hbar . \tag{3.52}$$

This inequality expresses the uncertainty principle (Sect. 2.6). We have thus verified the intimate connection between the commutation relation of two operators and the uncertainties in the measurement of the corresponding physical quantities.

The invariance with respect to the parity transformation $\hat{\Pi}$ ($x \to -x$) plays an important role in quantum mechanics. The fact that neither the kinetic energy, nor the harmonic oscillator potential energy, are altered by the parity transformation is expressed by the commutation relation

$$[\hat{H}, \hat{\Pi}] = 0 . \tag{3.53}$$

As a consequence of this relation, it is possible to know simultaneously the eigenvalues of the two operators $\hat{H}$, $\hat{\Pi}$ (see Sect. 2.6). In this case the eigenstates of the harmonic oscillator Hamiltonian are also eigenstates of the parity operator $\hat{\Pi}$. The eigenvalues of the operator $\hat{\Pi}$ are determined by the fact that the operator $\hat{\Pi}^2$ must have the single eigenvalue $\pi^2 = 1$, since the system is left unchanged after two applications of the parity transformation. There are thus two eigenvalues corresponding to the operator $\hat{\Pi}$, namely $\pi = \pm 1$. The eigenfunctions are either invariable under the parity transformation ($\pi = 1$, even functions) or change sign ($\pi = -1$, odd functions). This is verified in the case of the harmonic oscillator, since the operators a^+, a change sign under the parity transformation and the parity of the state labeled by the quantum number n is therefore

$$\hat{\Pi}\varphi_n = (-1)^n \varphi_n . \tag{3.54}$$

Problems

Problem 1. Consider the matrix

$$\begin{pmatrix} 0 & 1 & 0 \\ 1 & 0 & 1 \\ 0 & 1 & 0 \end{pmatrix} .$$

1. Find the eigenvalues and verify the conservation of the trace after diagonalization.
2. Find the eigenvector corresponding to each eigenvalue.
3. Check the orthogonality of states corresponding to different eigenvalues.
4. Construct the unitary transformation from the basic set of states used in (3.5) to the eigenstates of this matrix.

Problem 2. Consider the matrix

$$\begin{pmatrix} a & c \\ c & -a \end{pmatrix}.$$

1. Calculate the eigenvalues as a function of the real numbers a, c.
2. Show that the odd terms in c vanish in an expansion in powers of c ($|c| \ll |a|$).
3. Show that the linear term does not disappear if $|c| \gg |a|$.

Problem 3. Which of the following vector states are linearly independent?

$$\varphi_1 = \begin{pmatrix} i \\ 1 \end{pmatrix}, \quad \varphi_2 = \begin{pmatrix} -i \\ 1 \end{pmatrix}, \quad \varphi_3 = \begin{pmatrix} 1 \\ i \end{pmatrix}, \quad \varphi_4 = \begin{pmatrix} 1 \\ -i \end{pmatrix}.$$

Problem 4. Consider the two operators

$$\hat{Q} = \begin{pmatrix} 0.5 & 0 & 0 \\ 0 & 0.5 & 0 \\ 0 & 0 & -1 \end{pmatrix} \quad \text{and} \quad \hat{R} = \begin{pmatrix} 0 & 0.5 & 0 \\ 0.5 & 0 & 0 \\ 0 & 0 & 1 \end{pmatrix}.$$

1. Calculate the eigenvalues.
2. Determine whether or not the operators commute.
3. If so, obtain the simultaneous eigenvectors of both operators.

Problem 5. Consider a unit vector with components $\cos\beta$ and $\sin\beta$ along the z- and x-axes, respectively. The matrix representing the spin operator in this direction is written as $\hat{S}_\beta = \hat{S}_z \cos\beta + \hat{S}_x \sin\beta$.

1. Find the eigenvalues of $\hat{S}_\beta$ using symmetry properties.
2. Diagonalize the matrix.
3. Find the amplitudes of the new eigenstates in a basis for which the operator $\hat{S}_z$ is diagonal.

Problem 6. If a and a^+ are the annihilation and creation operators defined in (3.35), show that $[a, (a^+)^n] = n(a^+)^{(n-1)}$.

Problem 7.

1. Calculate the energy of a particle subject to the potential $V(x) = V_0 + c\hat{x}^2/2$ if the particle is in the third excited state.
2. Calculate the energy eigenvalues for a particle moving in the potential $V(x) = c\hat{x}^2/2 + b\hat{x}$.

Problem 8.

1. Express the distance x_c as a function of the mass M and the restoring parameter c used in Problem 7.
2. If c is multiplied by 9, what is the separation between consecutive eigenvalues?
3. Show that x_c is the maximum displacement of a classical particle moving in a harmonic oscillator potential with an energy of $\hbar\omega/2$.

Problem 9. Evaluate the matrix elements $\langle n+\eta|x^2|n\rangle$ and $\langle n+\eta|p^2|n\rangle$ in the harmonic oscillator basis, for $\eta = 1, 2, 3, 4$:

1. using the closure property and the matrix elements (3.47),
2. applying the operators $\hat{x}^2$ and $\hat{p}^2$, expressed in terms of the a^+, a, on the eigenstates (3.43).

Find the ratio $\langle n+\nu|K|n\rangle : \langle n+\nu|V|n\rangle$ ($\nu = 0, \pm 2$) in the harmonic oscillator basis, where $\hat{K}, \hat{V}$ are the operators corresponding to the kinetic and the potential energies, respectively. Justify the resulting sign difference between these three cases on quantum mechanical grounds.

Problem 10. Calculate the expectation value of the coordinate operator for a linear combination of harmonic oscillator states with the same parity.

Problem 11.

1. Construct the normalized, linear combination of harmonic oscillator states $\Psi = c_0\varphi_0 + c_1\varphi_1$ for which the expectation value $\langle \Psi|x|\Psi\rangle$ becomes maximized.
2. Evaluate in such a state the expectation values of the coordinate, the momentum and the parity operators.

Note: In some chemical bonds, nature takes advantage of the fact that electrons protrude from the atom in a state similar to the linear combination Ψ. This situation is called hybridization.

Problem 12. Verify the normalization of the states in (3.43).

4 The Schrödinger Realization of Quantum Mechanics

The realization of the basic principles of quantum mechanics by means of position wave functions is presented in Sect. 4.1. This is where the time-independent Schrödinger equation is obtained, and where the spatial dimension in quantum problems appears explicitly.

The harmonic oscillator problem is solved again in Sect. 4.2. The reader will thus be able to contrast these two realizations of quantum mechanics by comparing the results obtained here with those presented in Sect. 3.2.

Solutions to the Schrödinger equation in the absence of forces are discussed in Sect. 4.3. Such solutions present normalization problems which are solved by taking into consideration the limiting case of particles moving either in a large, infinitely deep square well potential, or along a circumference with a large radius (Sect. 4.4). These solutions are applied to some situations that are interesting both conceptually and in practical applications: the step potential (Sect. 4.5.1) and the square barrier (Sect. 4.5.2), which are schematic versions of scattering experiments. The free-particle solutions are also applied to the bound-state problem of the finite square well (Sect. 4.6*) and to the periodic potential (Sect. 4.7*).

4.1 Time-Independent Schrödinger Equation

In the formulation of quantum mechanics in this chapter, the state vector is a complex function of the coordinate, $\Psi = \Psi(x)$. This type of state vector is usually known as a wave function. The sum of two wave functions is another wave function:

$$\Psi(x) = \alpha_B \Psi_B(x) + \alpha_C \Psi_C(x) \, . \tag{4.1}$$

The scalar product is defined as

$$\langle \Psi_B | \Psi_C \rangle = \int_{-\infty}^{\infty} \Psi_B^* \Psi_C \, \mathrm{d}x \, . \tag{4.2}$$

As a consequence of this choice of $\Psi(x)$, the coordinate operator is simply the coordinate itself:

$$\hat{x} = x \, . \tag{4.3}$$

A realization of the algebra (2.8) is given by the assignment[1]

$$\hat{p} = -i\hbar \frac{d}{dx} \,, \qquad (4.4)$$

since for an arbitrary function $f(x)$,

$$[x, \hat{p}]f = -i\hbar x \frac{df}{dx} + i\hbar \frac{d(xf)}{dx} = i\hbar f \,. \qquad (4.5)$$

It is simple to verify that the operator x is Hermitian, according to the definition (2.34). This is also true for the momentum operator, since

$$\int_{-\infty}^{\infty} \Phi^* \hat{p} \Psi dx = -i\hbar \Phi^* \Psi \big|_{-\infty}^{\infty} + i\hbar \int_{-\infty}^{\infty} \Psi \frac{d}{dx} \Phi^* = \left(\int_{-\infty}^{\infty} \Psi^* \hat{p} \Phi dx \right)^* , \qquad (4.6)$$

where we have assumed $\Psi(\pm\infty) = 0$, as is the case for bound systems. The eigenfunctions of the momentum operator are discussed in Sect. 4.3.

A translation by the amount a can be performed by means of the unitary operator

$$\mathcal{U}(a) = \exp\left(\frac{i}{\hbar} a \hat{p} \right) , \qquad (4.7)$$

since

$$\mathcal{U}(a)\Psi(x) = \sum_n \frac{a^n}{n!} \frac{d^n \Psi}{dx^n} = \Psi(x + a) \,. \qquad (4.8)$$

A finite translation may be generated by a series of infinitesimal steps

$$\mathcal{U}(\delta a) = 1 + \frac{i}{\hbar} \delta a \, \hat{p} \,, \qquad (4.9)$$

and $\hat{p}$ is referred to as the generator of infinitesimal translations.

The replacement of the operators (4.3) and (4.4) in the classical expression of any physical observable $Q(x, p)$ yields the corresponding quantum mechanical operator $\hat{Q} = Q(x, \hat{p})$ in a differential form.[2] Given any complete set of orthonormal wave functions $\varphi_i(x)$, the matrix elements associated with the operator $\hat{Q}$ are constructed as in (2.6). Therefore this construction provides the link between the Heisenberg and Schrödinger realizations of quantum mechanics.

The Hamiltonian (2.9) yields the eigenvalue equation

$$-\frac{\hbar^2}{2M} \frac{d^2 \varphi_i}{dx^2} + V(x)\varphi_i = E_i \varphi_i \,, \qquad (4.10)$$

which is called the time-independent Schrödinger equation.

---

[1] Although any function of x may be added to (4.4) and still satisfy (2.8), such a term should be dropped because free space is homogeneous.

[2] There is an ambiguity if in the classical expression there appears a product of physical quantities QR corresponding to non-commuting operators. In such cases one uses the (Hermitian) average operator $(1/2)(\hat{Q}\hat{R} + \hat{R}\hat{Q})$ [see (2.32)].

4.1.1 Probabilistic Interpretation of Wave Functions

Information may be extracted from the wave function through the probability density (2.46) [27]

$$\rho(x) = |\Psi(x)|^2 \ . \tag{4.11}$$

The probability of finding the particle in the interval $L_1 \leq x \leq L_2$ is given by the integral

$$\int_{L_1}^{L_2} |\Psi(x)|^2 \mathrm{d}x \ . \tag{4.12}$$

In particular, the probability of finding the particle anywhere must equal one:

$$1 = \langle \Psi | \Psi \rangle \ , \tag{4.13}$$

which implies that the wave function should be normalized.

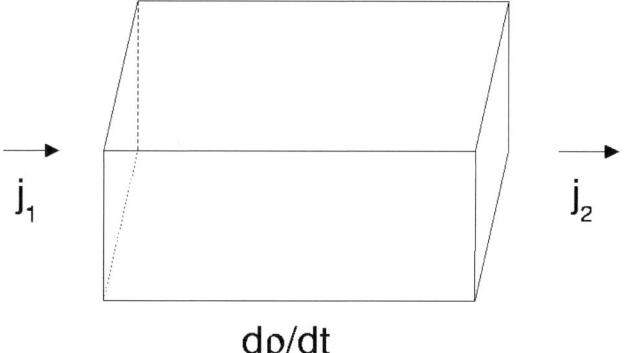

Fig. 4.1. Conservation of probability density. The rate of change within a certain interval is given by the flux differences at the boundaries of the interval

We now discuss how this probability changes with time t. We therefore allow for a time dependence of the wave function[3] $[\Psi = \Psi(x,t)]$:

$$\begin{aligned}\frac{\mathrm{d}}{\mathrm{d}t}\int_{L_1}^{L_2} |\Psi(x,t)|^2 \mathrm{d}x &= \int_{L_1}^{L_2} \left(\dot{\Psi}^*\Psi + \Psi^*\dot{\Psi}\right)\mathrm{d}x \\ &= \frac{\mathrm{i}}{\hbar}\int_{L_1}^{L_2} \left[\left(\mathrm{i}\hbar\dot{\Psi}\right)^*\Psi - \Psi^*\left(\mathrm{i}\hbar\dot{\Psi}\right)\right]\mathrm{d}x \ . \end{aligned} \tag{4.14}$$

We may replace $\mathrm{i}\hbar\dot{\Psi}$ with $\hat{H}\Psi$, according to the time-dependent Schrödinger equation (9.5). We are left with only the kinetic energy contribution, since the terms proportional to the potential cancel inside (4.14):

---

[3] The time dependence of the wave function is discussed in Chap. 9. We anticipate the result here because the notion of probability current is needed in the next few sections.

$$\frac{\mathrm{d}}{\mathrm{d}t}\int_{L_1}^{L_2}|\Psi(x,t)|^2\mathrm{d}x = -\frac{i\hbar}{2M}\int_{L_1}^{L_2}\left(\frac{\mathrm{d}^2\Psi^*}{\mathrm{d}x^2}\Psi - \Psi^*\frac{\mathrm{d}^2\Psi}{\mathrm{d}x^2}\right)\mathrm{d}x$$

$$= \frac{i\hbar}{2M}\int_{L_1}^{L_2}\frac{\mathrm{d}}{\mathrm{d}x}\left(-\frac{\mathrm{d}\Psi^*}{\mathrm{d}x}\Psi + \Psi^*\frac{\mathrm{d}\Psi}{\mathrm{d}x}\right). \quad (4.15)$$

We obtain the equation

$$\frac{\partial\rho}{\partial t} + \frac{\partial j}{\partial x} = 0, \quad (4.16)$$

where we have defined the probability current

$$j(x,t) \equiv -\frac{i\hbar}{2M}\left(-\frac{\partial\Psi^*}{\partial x}\Psi + \Psi^*\frac{\partial\Psi}{\partial x}\right). \quad (4.17)$$

Equation (4.16) is a continuity equation, similar to the one used in hydrodynamics to express conservation of mass. Imagine a long cylinder along the x-axis, bounded by two circles of area $\mathcal{A}$ at $x = L_1$ and $x = L_2$, respectively. The variation of the probability of finding the particle inside the cylinder, i.e.,

$$-\frac{\partial}{\partial t}\int_{L_1}^{L_2}\rho\,\mathrm{d}x,$$

is equal to the difference between the fluxes leaving and entering the cylinder, viz., $\mathcal{A}[j(L_2) - j(L_1)]$ (see Fig. 4.1).

The probability density and the probability current give spatial dimensions to the Schrödinger realization of quantum mechanics. These spatial features are especially useful in chemistry, where bulges of electron distribution in atoms are associated with increases in the chemical affinities of elements (Fig. 5.2).

The expression for the probability current underscores the need to use complex state vectors in quantum mechanics, since the current vanishes for real wave functions.

Here we may continue the list of misconceptions that prevail in quantum mechanics [21]:

- "The probability current $j(x)$ is related to the speed of that part of the particle which is located at the position x." This statement gives the false impression that, although the particle as a whole has neither a definite position nor a definite momentum, it is made up of parts that do. In fact, particles are not made up of parts.
- "For any energy eigenstate, the probability density must have the same symmetry as the Hamiltonian." This statement is correct in the case of inversion symmetry, for even states of a parity-invariant Hamiltonian (see Sect. 4.2). It is not correct for odd states. It is also generally false for a central potential, since only $l = 0$ states have probability densities with a spherical shape (Fig. 5.2).

- "A quantum state $\Psi(x)$ is completely specified by its associated probability density $|\Psi(x)|^2$." The probability densities, being real numbers, cannot give information about all the properties of the state, such as, for example, those related to momentum.
- "The wave function is dimensionless." It has the dimensions $[\text{length}]^{-dN/2}$, where N is the number of particles and d is the dimension of the space.
- "The wave function $\Psi(x)$ is a function of regular three-dimensional space." This is true only for one-particle systems. For two-particle systems, the wave function $\Psi(x_1, x_2)$ exists in six-dimensional, configuration space.
- "The wave function is similar to other waves appearing in classical physics." Unlike electromagnetic or sound waves, the wave function is an abstract entity. In particular, it does not interact with particles.

Both the probability density and the probability current are defined at each point in space. Other quantum predictions require integration over the whole space. For instance, the expectation value of an operator $\hat{Q}$ is

$$\langle \Psi | Q | \Psi \rangle \equiv \int_{-\infty}^{\infty} \Psi(x,t)^* \hat{Q} \Psi(x,t) \mathrm{d}x \;. \tag{4.18}$$

For an operator depending only on the coordinate x, this definition is a direct consequence of Born's probability density (4.11). However, for a differential operator such as $\hat{p}$, the alternative $\int \Psi(\hat{Q}\Psi)^* \mathrm{d}x$ is also possible. Nevertheless, the two definitions are identical for physical (Hermitian) operators.

4.2 The Harmonic Oscillator Revisited

The Schrödinger equation (4.10) corresponding to the harmonic oscillator Hamiltonian (3.30) reads

$$-\frac{\hbar^2}{2M} \frac{\mathrm{d}^2 \varphi_n(x)}{\mathrm{d}x^2} + \frac{M\omega^2}{2} x^2 \varphi_n(x) = E_n \varphi_n(x) \;. \tag{4.19}$$

4.2.1 Solution of the Schrödinger Equation

It is always useful to rewrite any equation in terms of dimensionless coordinates. Not only does one get rid of unnecessarily cumbersome constants, but the solution may apply just as well to cases other than the one being considered. Therefore, in the present problem, the coordinate x and the energy E are divided by the value of the characteristic length and energy (3.33), namely

$$u = x/x_c \;, \qquad e = E/\hbar\omega \;. \tag{4.20}$$

The Schrödinger equation thus simplified reads

$$-\frac{1}{2} \frac{\mathrm{d}^2 \varphi_n(u)}{\mathrm{d}u^2} + \frac{1}{2} u^2 \varphi_n(u) = e_n \varphi_n(u) \;. \tag{4.21}$$

This equation must be supplemented with the boundary conditions

$$\varphi_n(\pm\infty) = 0 . \tag{4.22}$$

The eigenfunctions and eigenvalues are of the form

$$\varphi_n(x) = N_n \exp\left(-\frac{1}{2}u^2\right) H_n(u) , \qquad e_n = n + \frac{1}{2} . \tag{4.23}$$

The H_n are Hermite polynomials[4] of degree $n = 0, 1, 2, \ldots$. The eigenfunctions and eigenvalues are also labeled by the quantum number n. Up to a phase, the constants N_n are obtained from the normalization condition (4.13)

$$N_n = 2^{n/2} \left(\pi^{\frac{1}{2}} n! x_c\right)^{-\frac{1}{2}} . \tag{4.24}$$

Since the Hamiltonian is a Hermitian operator, the eigenfunctions are orthogonal to each other and constitute a complete set of states:

$$\langle n|m\rangle = \int_{-\infty}^{\infty} \varphi_n^* \varphi_m \, dx = \delta_{nm} , \qquad \Psi(x) = \sum_n c_n \varphi_n . \tag{4.25}$$

The solutions corresponding to the lower quantum numbers are displayed in Table 4.1 and Fig. 4.2.

Table 4.1. Solutions to the harmonic oscillator problem for small values of n. P_n is defined in (4.28)

n	e_n	H_n	$N_n \pi^{1/4} x_c^{1/2}$	P_n [%]
0	1/2	1	1	15.7
1	3/2	u	$\sqrt{2}$	11.2
2	5/2	$u^2 - 1/2$	$\sqrt{2}$	9.5
5	11/2	$u^5 - 5u^3 + 15u/4$	$2/\sqrt{15}$	5.7

4.2.2 Spatial Features of the Solutions

The following features arise from the spatial dimension associated with the Schrödinger formulation:

---

[4] The reader is encouraged to verify that the few cases listed in Table 4.1 are correct solutions. Use can be made of the integrals

$$\int_{-\infty}^{\infty} \exp(-u^2) u^{2n} du = \frac{(2n-1)!!}{2^n} \pi^{\frac{1}{2}} , \qquad \int_{-\infty}^{\infty} \exp(-u^2) u^{2n+1} du = 0 .$$

1. Probability density. There are nodes in the probability density (except for the $n=0$ state). The existence of such nodes is incompatible with the classical notion of a trajectory $x(t)$, according to which the particle bounces from one side of the potential to the other, while going through every intermediate point. The fact is that the particle can never be found at the nodes. The quantum picture reminds us of the stationary wave patterns obtained, for instance, inside an organ pipe. The role that is played in the case of sound by the ends of the pipe, is played in the case of the quantum harmonic oscillator by the boundary conditions.
2. Comparison between the classical and quantum mechanical probability densities. The classical probability for finding a particle is inversely proportional to its speed $[v = (2E/M - \omega^2 x^2)^{1/2}$ for the harmonic oscillator]. Therefore, we may define a classical probability density $P_{\text{clas}} = \omega/\pi v$. The probability of finding the particle in any place, within the classically allowed interval $-x_n \leq x \leq x_n$, is one. Here, $x_n = x_c(2n+1)^{1/2}$ for a particle with energy $\hbar\omega(n + 1/2)$. The classical probability density displays a minimum around the origin and diverges as the particle approaches the end points of the allowed interval. The quantum mechanical density distribution for the ground state has exactly the opposite features. However, as n increases, the quantum mechanical density distribution tends[5] towards the classical limit (Fig. 4.2).
3. Tunnel effect. Outside the allowed interval $-x_c \leq x \leq x_c$, the classical particle would have a negative kinetic energy and thus an imaginary momentum. However, this argument does not hold in the quantum case, because it would imply some simultaneous determination of the particle location and the momentum, contradicting the uncertainty principle. Let us suppose that the particle has been detected in the interval $x_c \leq x \leq \sqrt{2}x_c$, i.e., within the region following the classically allowed one. In this interval, the probability density decreases from N_0^2/e at $x = x_c$ to N_0^2/e^3 (i.e., by a factor e^{-2}). If we measure the particle within this interval, and we take it to be a reasonable measure of the uncertainty in the position of the particle, then

$$\Delta x \approx 0.41 x_c . \tag{4.26}$$

According to the Heisenberg principle, the minimum uncertainty in the determination of the momentum is

$$\Delta p \geq 1.22\sqrt{\hbar\omega M} , \tag{4.27}$$

which is consistent with an uncertainty in the kinetic energy larger than $\Delta_p^2 \geq 1.5 M\hbar\omega$. Now, the potential energy in the same interval increases from $\hbar\omega/2$ to $\hbar\omega$. As a consequence, we cannot make any statement about a possible imaginary value for the momentum, which would rule out

---

[5] This is a manifestation of the correspondence principle, which was extensively used by Bohr in the old quantum theory.

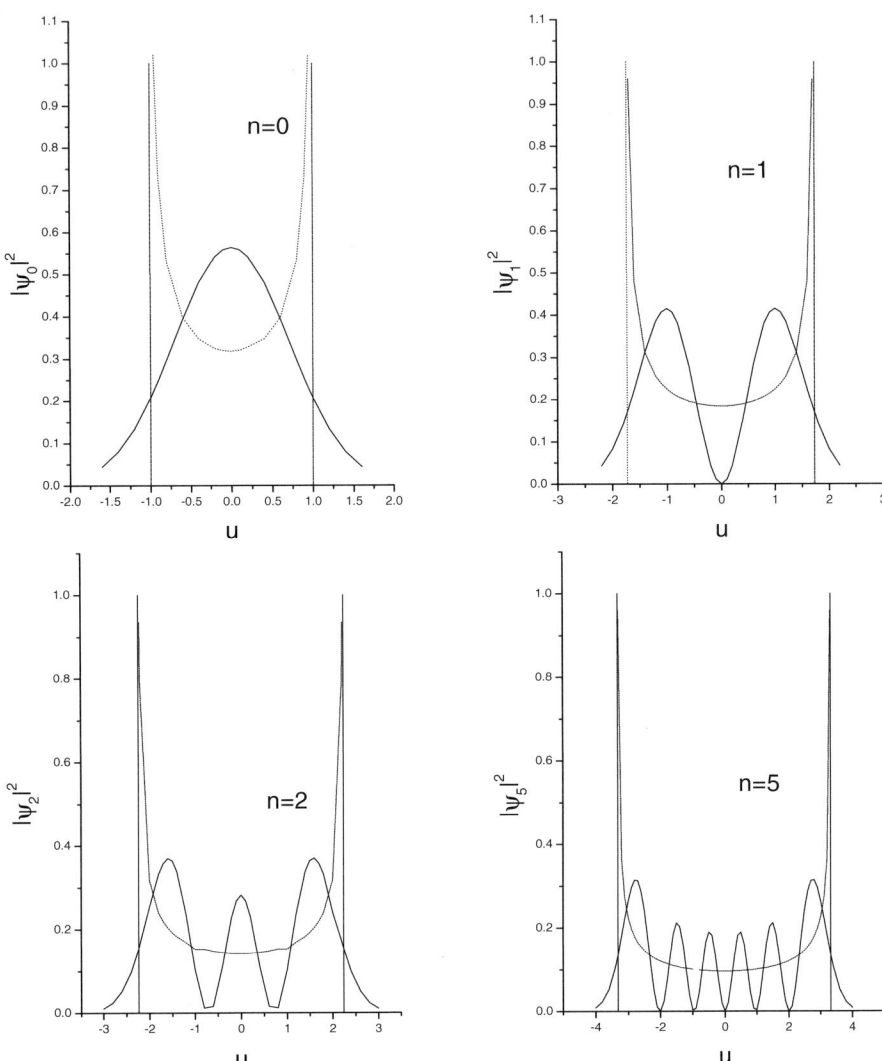

Fig. 4.2. Quantum mechanical probability densities and the classical probability densities of a harmonic oscillator potential as a function of the dimensionless distance u, for the quantum numbers $n = 0, 1, 2$ and 5. *Vertical lines* represent the classical amplitudes x_n

the possibility that the particle penetrates into the classically forbidden region.

The probability of finding the particle in the classically forbidden region is

$$P_n = \frac{2^{n+1}}{\pi^{1/2} n!} \int_{\sqrt{2n+1}}^{\infty} e^{-u^2} |H_n|^2 du \ . \tag{4.28}$$

This probability is a finite number, as large as 16% for the ground state. It decreases as the quantum number n increases, consistent with the tendency to approach the classical behavior for higher values of the energy (see Table 4.1).

4.3 Free Particle

If there are no forces acting on the particle, the potential is constant: $V(x) = V_0$. Let us assume in the first place that the energy $E \geq V_0$. In such a case the Schrödinger equation reads

$$-\frac{\hbar^2}{2M} \frac{d^2 \varphi_k(x)}{dx^2} = (E - V_0) \varphi_k(x) \ . \tag{4.29}$$

There are two independent solutions to this equation, namely

$$\varphi_{\pm k}(x) = A \exp(\pm i k x) \ , \qquad k = \frac{\sqrt{2M(E - V_0)}}{\hbar} \ . \tag{4.30}$$

The parameter k labeling the eigenfunction is called the wave number and has dimensions of a reciprocal length. The eigenvalues of the energy and the momentum are

$$E = \hbar^2 k^2 / 2M + V_0 \ , \qquad p = \hbar k \ . \tag{4.31}$$

Unlike the case of the harmonic oscillator (a typical bound case), the eigenvalues of both the momentum and the energy belong to a continuous set. The free-particle solutions satisfy the de Broglie relation [28]

$$p = \hbar k = h/\lambda \ . \tag{4.32}$$

The probability density is constant over the whole space

$$\rho_{\pm k}(x) = \varphi_{\pm k}^*(x) \varphi_{\pm k}(x) = |A|^2 \ , \tag{4.33}$$

while the probability current reads

$$j_{\pm k}(x) = -i \frac{\hbar}{2M} \left[\varphi_{\pm k}^* \frac{d\varphi_{\pm k}(x)}{dx} - \varphi_{\pm k}(x) \frac{d\varphi_{\pm k}^*(x)}{dx} \right] = \pm \frac{|A|^2 \hbar k}{M} \ . \tag{4.34}$$

These results pose normalization problems, which may be

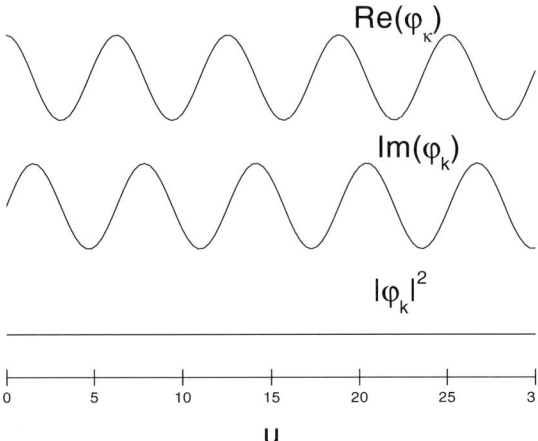

Fig. 4.3. Real component, imaginary component and modulus squared of a plane wave as functions of the dimensionless variable $u = kx$, where u is measured in radians

- solved by applying more advanced mathematical tools;
- taken care of through the use of tricks, as in Sect. 4.4;
- circumvented, by looking only at the ratios of the probabilities of finding the particle in different regions of space (Sects. 4.5.1 and 4.5.2).

Since there are two degenerate solutions,[6] the most general solution for a given energy E is a linear combination

$$\Psi(x) = A_+ \exp(ikx) + A_- \exp(-ikx) \,. \tag{4.35}$$

Let us consider now the case $E \leq V_0$, which makes no sense from the classical point of view. However, the solution of the harmonic oscillator problem (item 3 in Sect. 4.2.2) has warned us not to reject this situation out of hand in the quantum case. In fact, the general solution is the linear combination[7]

$$\Psi(x) = B_+ \exp(\kappa x) + B_- \exp(-\kappa x) \,, \qquad \kappa = -\mathrm{i}k = \frac{\sqrt{2M(V_0 - E)}}{\hbar} \,. \tag{4.36}$$

This general solution diverges at infinity: $|\Psi| \to \infty$ as $x \to \pm\infty$. Rather than a total rejection, this feature implies that the solution (4.36) can only be used if at least one of the extremes cannot be reached. For instance, if $V_0 > E$ for $x > a$, one imposes $B_+ = 0$.

---

[6] Two or more solutions are called degenerate if they are linearly independent and have the same energy.

[7] The only difference between the two solutions (4.35) and (4.36) is whether k is real or imaginary.

4.4 Infinite Square Well Potential. Electron Gas

The potential in this case is $V(x) = 0$ if $|x| \leq a/2$ and $V(x) = \infty$ for $|x| \geq a/2$ (Fig. 4.4).

The two infinite discontinuities should be canceled in the Schrödinger equation by similar discontinuities in the second derivative at the same points. This is accomplished by requiring the wave function to be a continuous function and requiring the first derivative to have a finite discontinuity at the boundaries of the potential. Since the wave function vanishes outside the classically allowed interval, the continuity of the wave function requires $\Psi(\pm a/2) = 0$.

According to (3.53), we may demand that the eigenfunctions of the Hamiltonian carry a definite parity. This is accomplished by using the solutions (4.35) with $A_+ = A_-$ for the even parity states, and $A_+ = -A_-$ for the odd ones. The eigenfunctions are written as

$$\varphi_n^{\text{even}}(x) = \sqrt{\frac{2}{a}} \cos(k_n x) , \qquad \varphi_n^{\text{odd}}(x) = \sqrt{\frac{2}{a}} \sin(k_n x) . \qquad (4.37)$$

inside the well and vanish outside the well. As a consequence of the boundary conditions,

$$\frac{k_n a}{2\pi} = n' , \qquad n' = \frac{1}{2}, 1, \frac{3}{2}, 2, \ldots , \qquad (4.38)$$

where the half-integer values correspond to the even solutions and the integer values to the odd ones. The eigenvalues of the energy are

$$E_n = \frac{\hbar^2 k_n^2}{2M} = \frac{\hbar^2 \pi^2}{2Ma^2} n^2 , \qquad (4.39)$$

with $n = 2n' = 1, 2, \ldots$.

The reader is recommended to check that the quantum features associated with the solution of the harmonic oscillator problem (Sect. 4.2.2) are reproduced in the case of the infinite square well. The exception is the one related to the tunnel effect, which is prevented here by the infinite discontinuity in the potential.

By increasing the size of the box, the infinite potential well may be used to model the potential binding the electrons in a metal. In the electron gas model in one dimension, the (non-interacting) electrons are confined to a (large) segment a, which is much larger than the size of a given experimental setup.

However, the standing waves (4.37) are not convenient for discussing charge and energy transport by electrons. In fact, the probability current associated with them vanishes. In the theory of metals, it is more convenient to use running waves $\exp(\pm ikx)$ (4.30). We may use an alternative boundary condition by imagining that the end point at $x = a/2$ is joined to the opposite point at $x = -a/2$. In this way the segment transforms into a circumference

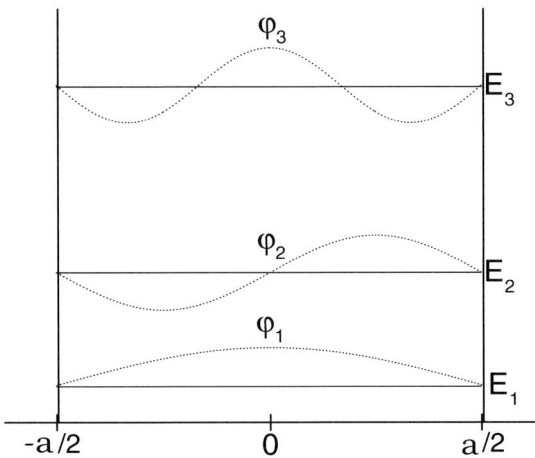

Fig. 4.4. Infinite square well potential. The energies E_n (*continuous lines*) and wave functions $\varphi_n(x)$ (*dotted curves*) are represented for the quantum numbers $n = 1, 2$ and 3

with the same length a. An electron arriving at the end of the well is not reflected back in, but leaves the metal and simultaneously re-enters at the opposite end. The representation of a free particle becomes more adequate as the radius $a/2\pi$ gets larger. This procedure results in the boundary condition $\Psi(x) = \Psi(x+a)$, or

$$\frac{k_n a}{2\pi} = n, \qquad n = 0, \pm 1, \pm 2, \ldots . \tag{4.40}$$

The total number of states is the same as for the standing waves (4.38), since the presence of half-integer and integer numbers in that case is compensated for here by the existence of two degenerate states $\pm n$. The eigenfunctions and energy eigenvalues are

$$\varphi_n(x) = \frac{1}{\sqrt{a}} \exp(\mathrm{i} k_n x), \qquad E_n = \frac{\hbar^2 k_n^2}{2M} . \tag{4.41}$$

As mentioned in Sect. 4.3, these functions are also eigenfunctions of the momentum operator $\hat{p}$ with eigenvalues $\hbar k_n$. Although the momenta (and the energies) are discretized, the gap

$$\Delta_k = 2\pi/a \tag{4.42}$$

between two consecutive eigenvalues becomes smaller than any prescribed interval, if the radius of the circumference is taken to be sufficiently large.

In quantum mechanics, sums over intermediate states often appear. In the case of wave functions of the type (4.41), this procedure may be simplified by transforming the sums into integrals, giving the length element in the integrals by $(a/2\pi)\mathrm{d}k$, according to (4.42):

$$\sum_k f_k \to \frac{a}{2\pi}\int f_k \mathrm{d}k \ . \tag{4.43}$$

The extension of the model in order to include a periodic crystal structure is performed in Sect. 4.7*. Calculations with the electron gas model for the three-dimensional case are carried out in Sect. 7.4.1.

4.5 One-Dimensional Unbound Problems

In this section we study problems related to the scattering of a particle by means of a potential. The particle impinges from the left and may be reflected and/or transmitted. There is no incoming wave from the right. Therefore the state vector must satisfy the following boundary conditions:

- it includes the term $A_+ \exp(\mathrm{i}kx)$ for $x \to -\infty$;
- it does not include the term $A_- \exp(-\mathrm{i}kx)$ for $x \to \infty$.

4.5.1 One-Step Potential

The one-step potential is written as $V(x) = 0$ for $x < 0$ and $V(x) = V_0 > 0$ for $x > 0$ (Fig. 4.5). It represents an electron moving along a conducting wire that is interrupted by a short gap. The electron feels a change in the potential as it crosses the gap.

$E_a < V_0$

Classically, the particle rebounds at $x = 0$ and cannot penetrate the region $x \geq 0$. Quantum mechanically this is no longer the case. For $x \leq 0$ the solution is given as the superposition of an incoming and a reflected wave (4.35), with $V_0 = 0$. Equation (4.36) holds for $x \geq 0$. This last solution cannot be rejected, since it does not diverge on the right half-axis if we impose the boundary condition $B_+ = 0$.

In order to have a Schrödinger equation valid at every point of space, the wave function and its first derivative should be continuous everywhere, including the point at which there is a finite discontinuity in the potential. These two requirements imply that

$$A_+ + A_- = B_- \ , \qquad A_+ - A_- = \mathrm{i}\frac{\kappa}{k}B_- \ . \tag{4.44}$$

Therefore

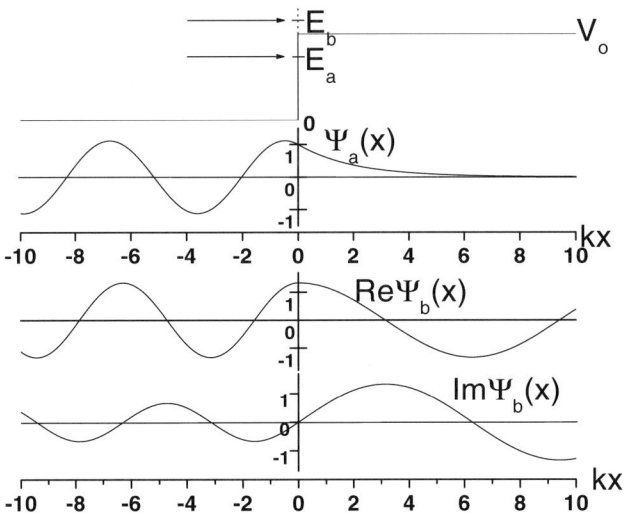

Fig. 4.5. One-step potential. Subscripts a and b label wave functions corresponding to energies $E_a = 3V_0/4$ and $E_b = 5V_0/4$, respectively

$$A_+ = \frac{1}{2}B_-\left(1 + i\frac{\kappa}{k}\right), \qquad A_- = \frac{1}{2}B_-\left(1 - i\frac{\kappa}{k}\right). \tag{4.45}$$

The total wave function is given by

$$\Psi_a(x) = \frac{1}{2}B_-\left[\left(1 + i\frac{\kappa}{k}\right)\exp(ikx) + \left(1 - i\frac{\kappa}{k}\right)\exp(-ikx)\right]$$
$$= B_-\left[\cos(kx) - \frac{\kappa}{k}\sin(kx)\right] \qquad (x \leq 0),$$
$$\Psi_a(x) = B_-\exp(-\kappa x) \qquad (x \geq 0). \tag{4.46}$$

The solution for $x \leq 0$ represents the superposition of an incident and a reflected wave. Since both amplitudes have equal moduli, they generate a standing wave with the corresponding nodes at positions such that $\tan(kx) = k/\kappa$, for $x \leq 0$ (see item 1 in Sect. 4.2.2). The probability currents associated with the incident and reflected waves are

$$j_\mathrm{I} = -j_\mathrm{R} = \frac{\hbar k}{M}\frac{|B_-|^2}{4}\left(1 + \frac{\kappa^2}{k^2}\right). \tag{4.47}$$

The reflection coefficient is defined as the absolute value of the ratio between reflected and incident currents. In the present case,

$$R \equiv \left|\frac{j_\mathrm{R}}{j_\mathrm{I}}\right| = 1. \tag{4.48}$$

The mutual cancellation between the two probability currents is correlated with the real character of the wave function (4.46).

There is a tunneling effect for $x \geq 0$, since the particle can penetrate into the forbidden region over a distance of the order of $\Delta x = 1/\kappa$. This length is accompanied by an uncertainty in the momentum and in the kinetic energy, so that

$$\Delta p \approx \frac{\hbar}{\Delta x} \approx \sqrt{2M(V_0 - E_a)}\,, \qquad \Delta E \approx \frac{(\Delta p)^2}{2M} \approx V_0 - E_a\,, \qquad (4.49)$$

respectively. The consequences of these uncertainties parallel those discussed in item 3 of Sect. 4.2.2.

$E_b > V_0$

The classical solution describes an incident particle which is totally transmitted, but with a smaller velocity. From the quantum mechanical point of view, the solution for $x \leq 0$ is again given by (4.35) with $V_0 = 0$, representing an incident plus a reflected wave. For $x \geq 0$ this same solution is valid, but with the wave number $k_b = \sqrt{2M(E_b - V_0)}/\hbar$. There is no incident wave from the right, since there is nothing that may bounce the particle back. Let C denote the amplitude of the transmitted wave $\exp(ik_b x)$. The continuity of the wave function and its first derivative at $x = 0$ requires that

$$A_+ + A_- = C\,, \qquad A_+ - A_- = \frac{k_b}{k} C\,. \qquad (4.50)$$

Using these equations, we may express the amplitudes of the reflected and transmitted waves as proportional to the amplitude of the incident wave, so that

$$\Psi_b(x) = \begin{cases} A_+ \left[\exp(ikx) + \dfrac{k - k_b}{k + k_b} \exp(-ikx) \right] & (x \leq 0)\,, \\ A_+ \dfrac{2k}{k + k_b} \exp(ik_b x) & (x \geq 0)\,. \end{cases} \qquad (4.51)$$

The probability currents associated with the incident, reflected and transmitted waves are

$$\begin{aligned} j_\mathrm{I} &= \frac{\hbar k}{M} |A_+|^2\,, \\ j_\mathrm{R} &= -\frac{\hbar k}{M} \left(\frac{k - k_b}{k + k_b} \right)^2 |A_+|^2\,, \\ j_\mathrm{T} &= \frac{\hbar k_b}{M} \left(\frac{2k}{k + k_b} \right)^2 |A_+|^2\,, \end{aligned} \qquad (4.52)$$

respectively. In this case we also define a transmission coefficient $T \equiv j_\mathrm{T}/j_\mathrm{I}$

$$R = \left(\frac{k - k_b}{k + k_b} \right)^2\,, \qquad T = \frac{4k k_b}{(k + k_b)^2}\,, \qquad (4.53)$$

54   4 The Schrödinger Realization of Quantum Mechanics

and we find that $R+T=1$ as expected, since the current should be conserved in the present case.

What makes the particle bounce? The quantum mechanical situation is similar to a beam of light crossing the boundary between two media with different indices of refraction. At least a partial reflection of the beam takes place.

Note that the wave functions (4.46) and (4.51) may be obtained from each other through the substitution $k_b(E) = i\kappa(E)$.

4.5.2 Square Barrier

The potential is given by $V(x) = 0$ ($|x| > a/2$) and $V(x) = V_0$ ($|x| < a/2$) (Fig. 4.6). We only consider explicitly the case $E \leq V_0$. Classically, the particle can only be reflected at $x = -a/2$.

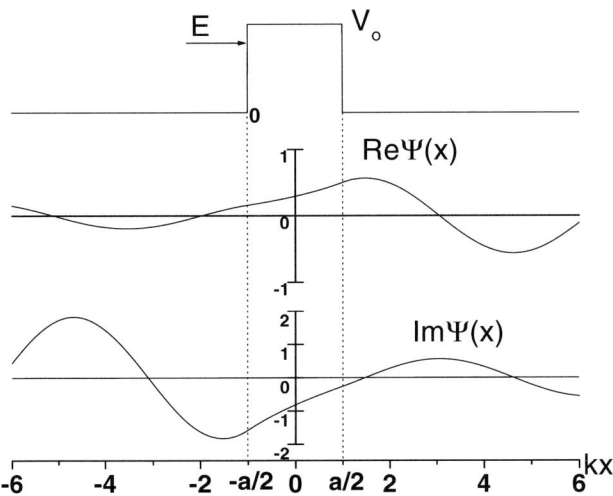

Fig. 4.6. Square barrier and associated wave function. Here $E = 3V_0/4$

For $x \leq -a/2$ and for $x \geq a/2$, the solution to the Schrödinger equation again takes the form (4.35), with the same value of k for both regions ($V_0 = 0$). However, there is only a transmitted wave, $C \exp(ikx)$, for $x \geq a/2$. Within the intermediate region $-a/2 \leq x \leq a/2$, the solution is as in (4.36). We cannot now reject either of the two components on account of their bad behavior at infinity. We thus have five amplitudes. The continuity conditions at the two boundaries provide us with four equations: the four remaining amplitudes may be expressed in terms of the amplitude of the incident wave A_+. We may also obtain here the currents associated with the incident beam j_I, the reflected beam j_R, the transmitted beam j_T and the beam within the barrier j_B, and the reflection and transmission coefficients R, T:

$$j_\mathrm{I} = \frac{\hbar k}{M}|A_+|^2, \qquad j_\mathrm{R} = -\frac{\hbar k}{M}|A_-|^2,$$

$$j_\mathrm{T} = \frac{\hbar k}{M}|C|^2, \qquad j_\mathrm{B} = \frac{2\hbar\kappa}{M}\left[\mathrm{Re}\,(B_+)\mathrm{Im}\,(B_-) - \mathrm{Re}\,(B_-)\mathrm{Im}\,(B_+)\right],$$

$$R = \left|\frac{j_\mathrm{R}}{j_\mathrm{I}}\right| = \frac{\sinh^2(\kappa a)}{\dfrac{4E}{V_0}\left(1 - \dfrac{E}{V_0}\right) + \sinh^2(\kappa a)},$$

$$T = \left|\frac{j_\mathrm{T}}{j_\mathrm{I}}\right| = \left|\frac{j_\mathrm{B}}{j_\mathrm{I}}\right| = \frac{\dfrac{4E}{V_0}\left(1 - \dfrac{E}{V_0}\right)}{\dfrac{4E}{V_0}\left(1 - \dfrac{E}{V_0}\right) + \sinh^2(\kappa a)}. \tag{4.54}$$

Transmission through a potential barrier is another manifestation of the tunnel effect, which has been discussed both in connection with the harmonic oscillator (item 3 in Sect. 4.2.2) and with the one-step potential (Sect. 4.5.1). The tunnel effect manifests itself in the α-decay of nuclei, the tunneling microscope, etc.

4.6* Finite Square Well Potential

The potential reads $V(x) = -V_0 < 0$ if $|x| < a/2$ and $V(x) = 0$ for other values of x. Here we consider only bound states, with a negative energy ($0 \geq -E \geq -V_0$).

As in the harmonic oscillator case, the potential is invariant under the parity transformation $x \to -x$. Thus we expect the eigenfunctions to be either even or odd with respect to this transformation. Therefore, the solution (4.35) applies in the region $|x| \leq a/2$, with

$$A_+ = \pm A_- \quad \text{and} \quad k = \frac{1}{\hbar}\left[2M(V_0 - E)\right]^{1/2}.$$

Moreover, invariance under the parity transformation allows one to confine calculation of the boundary conditions to the position $x = a/2$. The wave function to the right of this point is given by (4.36) with $B_+ = 0$ and $\kappa = (1/\hbar)(2ME)^{1/2}$.

The ratio between the continuity conditions corresponding to the first derivative and to the function itself yields the eigenvalue equation

$$\frac{\kappa}{k} = \tan\frac{ka}{2}. \tag{4.55}$$

This is as far as we can go analytically in this case. Equation (4.55) must either be solved numerically or using the following graphical method: the equation determining the value of k is equivalent to the equation $E = V_0 - \hbar^2 k^2/2M$. Therefore, we obtain the ratio

$$\frac{\kappa}{k} = \sqrt{\frac{2MV_0}{\hbar^2 k^2} - 1} \,, \tag{4.56}$$

and (4.55) becomes

$$\sqrt{\frac{MV_0 a^2}{2\hbar^2 \theta^2} - 1} = \tan \theta \,, \tag{4.57}$$

where $\theta \equiv ka/2$. The function $\tan \theta$ increases from zero to infinity in the interval $0 \leq \theta \leq \pi/2$, while the left-hand side decreases from infinity to a finite value as θ increases in the same interval. Therefore, there is a value of θ at which the two curves intersect, corresponding to the lowest eigenvalue. An analogous argument is made for the successive roots of (4.57). The nth root is found in the interval $(n-1)\pi \leq \theta \leq (n-1/2)\pi$.

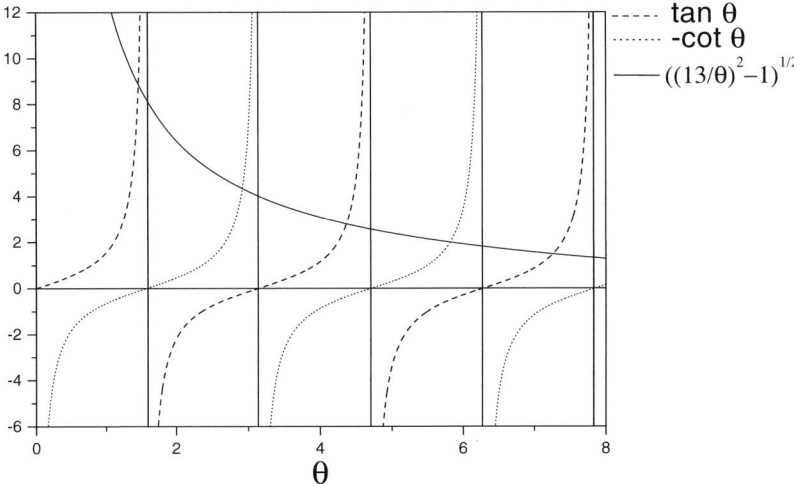

Fig. 4.7. Graphical determination of the energy eigenvalues of a finite square well potential. The intersections of the *continuous curve* $\sqrt{(13/\theta)^2 - 1}$ with the *dashed curves* correspond to even-parity solutions, whilst those with the *dotted curves* correspond to the odd ones

Unlike the harmonic oscillator case, the number of roots is limited, since (4.57) requires $\theta \leq \theta_{\max}$, where

$$\theta_{\max} = \sqrt{\frac{MV_0 a^2}{2\hbar^2}} \,. \tag{4.58}$$

There is a set of odd solutions that satisfy an equation similar to (4.55), namely

$$-\cot \frac{ka}{2} = \frac{\kappa}{k} \,. \tag{4.59}$$

Unlike the classical case, the probability density is not constant in the interval $|x| \leq a/2$. Moreover, there is a finite probability of finding the particle outside the classically allowed region. However, the solutions tend towards the classical behavior as n increases.

The spectrum of normalizable (bound) states is always discrete. Conversely, states that have a finite amplitude at infinity must be part of a continuous spectrum. This is the case for positive values of the energy. The corresponding analysis parallels the one already made for the square barrier, with the difference that a solution of type (4.35), instead of (4.36), should also be used for the region inside the well. There will also be incident, reflected and transmitted waves, and coefficients of reflection and transmission that sum to a value of unity.

4.7* Band Structure of Crystals

A crystal consists of an array of N positive ions displaying a periodic structure in space and electrons moving in the electric field generated by the ions. Figure 4.8 sketches the potential $V(x+d) = V(x)$ that an electron feels in the one-dimensional case. In this section we study the main features of the single-particle eigenstates in such a potential.

Classically, an electron moving in the potential of Fig. 4.8 may be bound to a single ion so that it is unable to transfer to another ion. Quantum mechanically, this may be ensured only if the distance d between the ions is very large. In such a case, the N states in which the electron is bound to one atom of the array constitute an orthogonal set of states which is N times degenerate. However, as the distance d is reduced to realistic values, we expect the degeneracy to be broken and the energy eigenvalues to be distributed within a band. In the following, we show how this picture is represented mathematically.

The Bloch theorem states that the wave function of a particle moving in a periodic potential has the form

$$\varphi_k(x) = \exp(ikx) u_k(x) , \qquad (4.60)$$

where k is real and $u_k(x+d) = u_k(x)$ is a periodic function [29].

Since the most relevant property of the potential is its periodicity and not its detailed shape, we replace the realistic potential with a periodic array of square well potentials (Fig. 4.8). We have learned by now how to solve square well problems. Here $V(x) = -V_0$ ($0 \leq x \leq b$) and $V(x) = -V_1$ ($b \leq x \leq d$). Moreover, $V(x+d) = V(x)$. We denote the energy by $-E$, with $E > 0$. We assume that the electron is bound to the crystal for negative energy values.

Region I: $-V_0 \leq -E \leq -V_1$

According to Sect. 4.3, the wave functions in the interval $nd \leq x \leq (n+1)d$ are

58   4 The Schrödinger Realization of Quantum Mechanics

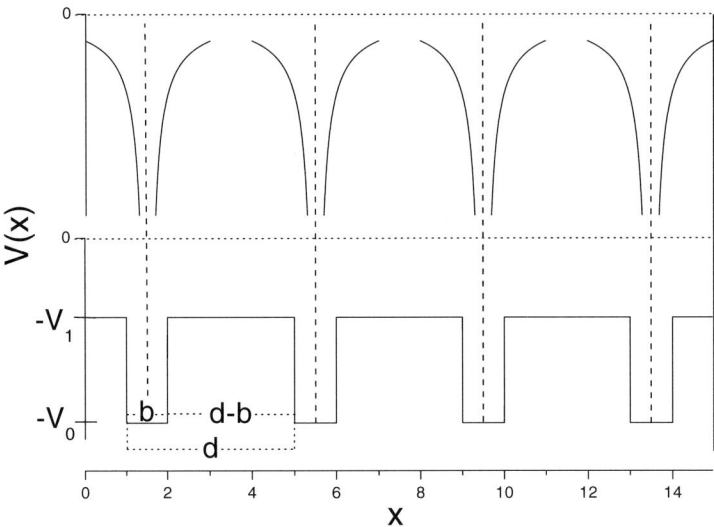

Fig. 4.8. Periodic potential

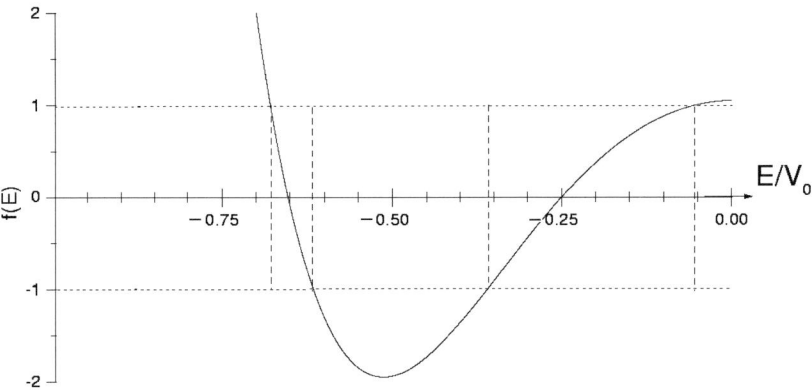

Fig. 4.9. Available intervals of energy (bands), obtained with the periodic square potential of Fig. 4.8

$$\Psi(x) = \begin{cases} A_+ \exp(ik_b x) + A_- \exp(-ik_b x) \,, & nd \leq x \leq nd+b \,, \\ B_+ \exp(\kappa_b x) + B_- \exp(-\kappa_b x) \,, & nd+b \leq x \leq (n+1)d \,, \end{cases} \tag{4.61}$$

where

$$k_b = \frac{1}{\hbar}\sqrt{2M(V_0 - E)} \,, \qquad \kappa_b = \frac{1}{\hbar}\sqrt{2M(E - V_1)} \,. \tag{4.62}$$

Thus the periodic function $u_k(x)$ is of the form

$$u_k(x) = \begin{cases} A_+ \exp[\mathrm{i}(k_b - k)x] + A_- \exp[-\mathrm{i}(k_b + k)x] \,, & nd \leq x \leq nd + b \,, \\ B_+ \exp[(\kappa_b - \mathrm{i}k)x] + B_- \exp[-(\kappa_b + \mathrm{i}k)x] \,, & \\ & nd + b \leq x \leq (n+1)d \,. \end{cases} \quad (4.63)$$

The periodicity of u_k requires

$$u_k(x) = A_+ \exp[\mathrm{i}(k_b - k)(x - d)] + A_- \exp[-\mathrm{i}(k_b + k)(x - d)] \,, \quad (4.64)$$

for

$$(n+1)d \leq x \leq (n+1)d + b \,.$$

The continuity conditions for the wave function, or equivalently for u_k, yield four linear equations for the amplitudes $A_\pm, B_\pm$ (two at $x = b$ and two at $x = d$). Therefore the determinant of the coefficients of the amplitudes should vanish. This condition leads to the equation

$$f(E) = \cos(kd) \,, \quad (4.65)$$

where

$$f(E) = \frac{\kappa_b^2 - k_b^2}{2k_b \kappa_b} \sinh\left[\kappa_b(d - b)\right] \sin(k_b b) + \cosh\left[\kappa_b(d - b)\right] \cos(k_b b) \,. \quad (4.66)$$

Region II: $-V_1 \leq -E \leq 0$

The procedure is completely parallel to the previous case except for the fact that the wave function is also of the form (4.35) in the interatomic space $b \leq x \leq d$, with $k_c = (1/\hbar)\sqrt{2M(E - V_1)}$. Equation (4.65) still holds, with

$$f(E) = -\frac{k_c^2 + k_b^2}{2k_b k_c} \sin\left[k_c(b - d)\right] \sin(k_b b) + \cos\left[k_c(d - b)\right] \cos(k_b b) \,. \quad (4.67)$$

The allowed values of E fall into bands satisfying the condition $|f(E)| \leq 1$. Figure 4.9 represents the function $f(E)$, encompassing the two regions I and II, for the parameters $V_1 = V_0/2$ and $b = \hbar\sqrt{2/MV_0}$.

Equation (4.65) remains unchanged if k is increased by a multiple of $2\pi/d$. We therefore confine k to the interval

$$-\frac{\pi}{d} \leq k \leq \frac{\pi}{d} \,. \quad (4.68)$$

We now apply the periodic boundary conditions discussed in Sect. 4.4. The length of the circumference is $a = Nd$. Therefore,

$$\exp(\mathrm{i}k_n N d) = 1 \,, \quad k_n = \pm\frac{2\pi n}{Nd} \,, \quad n = 0, \pm 1, \pm 2, \pm\frac{1}{2}N \,, \quad (4.69)$$

where the limits (4.68) have been taken into account. There are as many possible values of k as there are ions in the array. This result is consistent with the fact that binding the electron to each ion also constitutes a possible solution to the problem, as mentioned at the beginning of this section.

Problems

Problem 1. Show that the definition (4.18) of the expectation value coincides with the definition (2.18), if one uses the closure property appropriate for the position-dependent wave functions: $\sum_i \varphi_i(x)\varphi_i^*(y) = \delta(x-y)$.

Problem 2. Using Table 4.1, verify that:

1. the operator a (3.35) annihilates the ground state wave function φ_0;
2. the operator a^+, applied to $\varphi_1(x)$ yields $\sqrt{2}\varphi_2(x)$.

Hint: express the operators a, a^+ as differential operators.

Problem 3. Assume an infinite square well such that $V(x) = 0$ in the interval $0 < x < a$ and $V(x) = \infty$ for the remaining values of x.

1. Calculate the energies and wave functions.
2. Compare these results with those obtained in the text centering the well at the origin and explain the agreement on physical grounds.
3. Do the wave functions obtained in the first part have a definite parity?

Problem 4. Relate the minimum energy for a particle moving in a square well to the Heisenberg uncertainty principle.

Problem 5. Find the eigenvalue equations for a particle moving in a potential well such that $V(x) = \infty$ for $|x| \geq a/2$, $V(x) = V_0 \geq 0$ for $-a/2 \leq x \leq 0$, and $V(x) = 0$ for $0 < x < a/2$. Assume $0 \leq E \leq V_0$.

Problem 6. Estimate the error if we use (4.43) in the calculation of $\sum_k E_k$. Hint: recall that

$$\sum_{n=0}^{n=\nu} n^2 = \frac{\nu}{6}(\nu+1)(2\nu+1).$$

Problem 7. Calculate the transmission and reflection coefficients for an electron with a kinetic energy of $E = 2$ eV coming from the right. The potential is $V(x) = 0$ for $x \leq 0$ and $V(x) = V_0 = 1$ eV for $x \geq 0$.

Problem 8. The highest energy of an electron inside a block of metal is 5 eV (Fermi energy). The additional energy that is necessary to remove the electron from the metal is 3 eV (work function).

1. Estimate the distance through which the electron penetrates the barrier, assuming that the width of the (square) barrier is much greater than the penetration distance.
2. Estimate the transmission coefficient if the width of the barrier is 20 Å.

Problem 9. Obtain the transmission coefficients of a potential barrier in the limits $\kappa a \ll 1$ and $\kappa a \gg 1$.

Problem 10. Consider a square well such that $V(x) = \infty$ for $x < 0$, $V(x) = 0$ for $0 < x < a/2$ and $V(x) = V_0$ for $x > a/2$.

1. Write down the equation for the eigenvalues.
2. Compare this equation with the one obtained for the finite square well in Sect. 4.6*.
3. For $V_0 \to \infty$, show that the wave function for the finite well satisfies the condition that it vanishes at $x = a/2$ and does not penetrate the classically forbidden region.

Problem 11. Calculate the number of even-parity states (EPS) and odd parity states (OPS) for a finite square well potential of depth V_0 centered at the origin, if the parameter

$$\theta = \frac{a}{\hbar}\sqrt{\frac{MV_0}{2}}$$

lies in the intervals: $(0, \pi/2)$, $(0, \pi)$, $(0, 3\pi/2)$, and $(0, 2\pi)$.

Problem 12.

1. Show that the eigenfunction of the Hamiltonian of a periodic potential is not an eigenfunction of the momentum operator.
2. Why is it not a momentum eigenstate?
3. Give an expression for the expectation value of the momentum.

5 Angular Momentum

This chapter and the next are devoted to single-particle problems in three dimensions. Chapter 5 treats only the kinematical aspects of these problems, which include the introduction of the most important quantum mechanical observable: the spin.

The commutation relation (2.8) is straightforwardly generalized to the three-dimensional case

$$[\hat{x}_i, \hat{p}_j] = i\hbar \delta_{ij} \ . \tag{5.1}$$

In classical physics, angular momentum is a physical, observable vector $\boldsymbol{L}$ that plays an important role, since it is a conserved quantity in the absence of external torques $\boldsymbol{\tau}$:

$$\boldsymbol{L} = \boldsymbol{r} \times \boldsymbol{p} \ , \qquad \frac{\mathrm{d}\boldsymbol{L}}{\mathrm{d}t} = \boldsymbol{\tau} \ . \tag{5.2}$$

As in the case of the Schrödinger equation, we quantize the problem by substituting

$$\hat{p}_i \to -i\hbar \frac{\partial}{\partial x_i} \tag{5.3}$$

into the classical expression (5.2). One obtains the commutation relations

$$[\hat{L}_x, \hat{L}_y] = i\hbar \hat{L}_z \ , \qquad [\hat{L}_y, \hat{L}_z] = i\hbar \hat{L}_x \ , \qquad [\hat{L}_z, \hat{L}_x] = i\hbar \hat{L}_y \ , \tag{5.4}$$

$$[\hat{L}^2, \hat{L}_x] = [\hat{L}^2, \hat{L}_y] = [\hat{L}^2, \hat{L}_z] = 0 \ . \tag{5.5}$$

5.1 Eigenvalues and Eigenstates

5.1.1 Matrix Treatment

In the following, we take the commutation relations (5.4) as the definition of quantum angular momentum. Therefore, this definition also takes care of the quantum version of orbital angular momentum (5.2). However, as we shall see, definition (5.4) also includes other types of angular momenta of a purely quantum mechanical origin. From here on we let $\hat{J}_i$ denote operator components that satisfy the relations

$$[\hat{J}_i, \hat{J}_j] = i\hbar \epsilon_{ijk} \hat{J}_k \ , \tag{5.6}$$

where ϵ_{ijk} is the Levi–Civita tensor[1], whatever their origin may be. We use the notation $\hat{L}_i$ for angular momentum operators associated with orbital motion (5.2).

The commutation relations ensure that one can precisely determine the modulus squared simultaneously with one projection of the angular momentum, but not two projections at the same time. Consequently, one may construct eigenfunctions that are common to the operators $\hat{J}^2$ and $\hat{J}_z$. The choice of the z-component is arbitrary, since the space is isotropic and, consequently, there are no preferred directions.

The procedure for solving this problem closely follows the matrix treatment of the harmonic oscillator (Sect. 3.2.1). It is given in detail in Sect. 5.4*. The following results are obtained:

- The eigenvalue equations for the operators $\hat{J}^2$ and $\hat{J}_z$ can be written as

$$\hat{J}^2 \varphi_{jm} = \hbar^2 j(j+1) \varphi_{jm}, \qquad \hat{J}_z \varphi_{jm} = \hbar m \varphi_{jm}, \qquad (5.7)$$

where the possible values of the quantum numbers j, m are

$$-j \leq m \leq j, \qquad j = 0, \frac{1}{2}, 1, \frac{3}{2}, \ldots, \qquad (5.8)$$

with m increasing in units of one.

- Since the maximum value of m is j, and $j^2 < j(j+1)$, the maximum projection of the angular momentum is always smaller than the modulus (except for $j = 0$). Thus, the angular momentum vector can never be completely aligned with the z-axis. This fact is consistent with the lack of commutativity in (5.6): a complete alignment would imply the vanishing of the components $\hat{J}_x, \hat{J}_y$ and thus the simultaneous determination of the corresponding physical quantities and of J_z (see Problem 3).

Figure 5.1 represents the possible orientations of the angular momentum vector for the case $j = 5/2$. It looks as if the angular momentum precesses around the z-axis. However, this picture is incorrect, since it implies that the end point of the angular momentum vector goes through a circular trajectory, something that does not make sense from the point of view of quantum uncertainty relations.

- The operators $\hat{J}_x$ and $\hat{J}_y$ display non-diagonal matrix elements within the basis (5.7), namely,

$$\langle j'm'|J_x|jm\rangle = \delta_{j'j}\delta_{m'(m\pm 1)}\frac{\hbar}{2}\sqrt{(j\mp m)(j\pm m+1)},$$

$$\langle j'm'|J_y|jm\rangle = \mp\delta_{j'j}\delta_{m'(m\pm 1)}\frac{i\hbar}{2}\sqrt{(j\mp m)(j\pm m+1)}. \qquad (5.9)$$

- None of the operators $\hat{J}_x, \hat{J}_y, \hat{J}_z, \hat{J}^2$ connect states with different values of the quantum number j.

---

[1] $\epsilon_{ijk} = 1$ if i, j, k are cyclical (as for $i = z, j = x, k = y$); otherwise $\epsilon_{ijk} = -1$ (as for $i = z, j = y, k = x$).

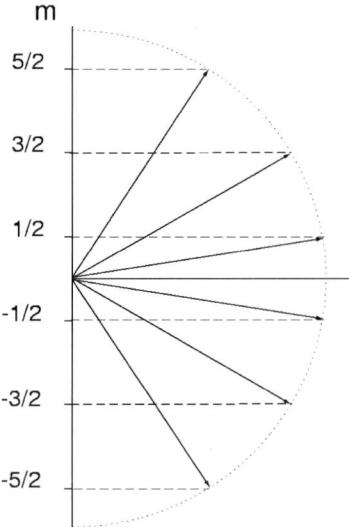

Fig. 5.1. Possible orientations of a $j = 5/2$ angular momentum vector

- In analogy with (4.7), the unitary operator associated with rotations is

$$\mathcal{U}(\boldsymbol{\alpha}) = \exp\left(\frac{\mathrm{i}}{\hbar}\boldsymbol{\alpha}\cdot\hat{\boldsymbol{J}}\right) . \qquad (5.10)$$

The rotation is specified by the axis of rotation (direction of the vector $\boldsymbol{\alpha}$) and the magnitude of the rotation angle α. The operator $\hat{J}_i$ is referred to as the generator of rotations around the i-axis.

5.1.2 Treatment Using Position Wave Functions

The concept of orbital angular momentum is especially useful in problems with spherical symmetry (like those involving atoms, nuclei, etc.), for which it is convenient to use the spherical coordinates

$$x = r\sin\theta\cos\phi , \qquad y = r\sin\theta\sin\phi , \qquad z = r\cos\theta ,$$

$$\mathrm{d}x\,\mathrm{d}y\,\mathrm{d}z = r^2 \sin\theta\,\mathrm{d}r\,\mathrm{d}\theta\,\mathrm{d}\phi , \qquad (5.11)$$

$$0 \leq \theta \leq \pi , \qquad 0 \leq \phi \leq 2\pi , \qquad 0 \leq r \leq \infty .$$

In these coordinates, the orbital angular momentum operators read

$$\hat{L}_x = \mathrm{i}\hbar\left(\sin\phi\frac{\partial}{\partial\theta} + \cot\theta\cos\phi\frac{\partial}{\partial\phi}\right) ,$$

$$\hat{L}_y = \mathrm{i}\hbar\left(-\cos\phi\frac{\partial}{\partial\theta} + \cot\theta\sin\phi\frac{\partial}{\partial\phi}\right) ,$$

$$\hat{L}_z = -i\hbar \frac{\partial}{\partial \phi},$$

$$\hat{L}^2 = -\hbar^2 \left(\frac{\partial^2}{\partial \theta^2} + \cot\theta \frac{\partial}{\partial \theta} + \frac{1}{\sin^2\theta} \frac{\partial^2}{\partial \phi^2} \right). \tag{5.12}$$

The detailed treatment of the orbital angular momentum operator is given in Sect. 5.5*. The results of such an approach are as follows:

- The simultaneous eigenfunctions of the operators $\hat{L}^2, \hat{L}_z$ are called spherical harmonics and denoted by $Y_{lm_l}(\theta,\phi)$. They satisfy the eigenvalue equations

$$\hat{L}_z Y_{lm_l} = \hbar m_l Y_{lm_l},$$
$$\hat{L}^2 Y_{lm_l} = \hbar^2 l(l+1) Y_{lm_l}, \tag{5.13}$$
$$\hat{\Pi} Y_{lm_l} = (-1)^l Y_{lm_l},$$

where

$$-l \leq m_l \leq l, \qquad l = 0, 1, 2, \ldots. \tag{5.14}$$

and $\hat{\Pi}$ is the parity operator[2] (3.54).
- Using the expressions (5.12), one may construct the matrix elements of the operators $\hat{L}_x, \hat{L}_y$. One obtains the same form as the matrix elements in (5.9), with the replacement $j \to l$, $m \to m_l$.
- The spherical harmonics constitute a complete set of single-valued basis states on the surface of a sphere of unit radius:

$$\Psi(\theta,\phi) = \sum_{lm_l} c_{lm_l} Y_{lm_l}. \tag{5.15}$$

- Figure 5.2 displays the projection of some spherical harmonics on the (x,z) plane. The protruding shapes have important consequences in the construction of chemical bonds.
- The rotational Hamiltonian of a molecule is proportional to the operator $\hat{L}^2$. The corresponding energy eigenvalues therefore follow the rule $l(l+1)$ (see Sect. 8.4.2).

By taking the commutation relations as the definition of the angular momentum operators, we have obtained operators that are not derived from the classical orbital angular momentum (see Sects. 5.1.1 and 5.1.2). This statement is supported by the fact that the quantum numbers j, m associated with these quantum-mechanical angular momenta may take either integer or half-integer values, in contrast with those labeling the orbital angular momentum, which can only take integer values. Otherwise we obtain the same

---

[2] For the three-dimensional case, the parity operation is written as $\boldsymbol{r} \to -\boldsymbol{r}$ or equivalently, $r \to r$, $\theta \to \pi - \theta$, $\phi \to \pi + \phi$.

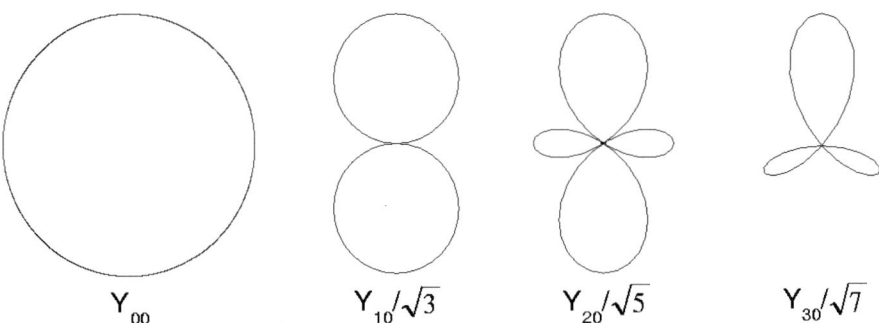

Fig. 5.2. Projection of the spherical harmonics Y_{l0} on the (x,z) plane, for the values $l = 0$–3. While the Y_{l0} are axially symmetric wave functions, $l = 0$ implies full spherical symmetry. The distance from the center to the top of the figures is $1/\sqrt{4\pi}$.

matrix elements for the projections $\hat{J}_i$ (5.9) as for the orbital angular momentum projections $\hat{L}_i$. On the other hand, the probability densities associated with the orbital angular momentum display interesting and useful features that are lacking in the more general derivation (Fig. 5.2).

5.2 Spin

5.2.1 Stern–Gerlach Experiment

A particle with a magnetic moment $\boldsymbol{\mu}$ and subject to a magnetic field $\boldsymbol{B}$ experiences a torque $\boldsymbol{\tau}$. When the particle is rotated through an angle $d\theta$ about the direction of $\boldsymbol{\tau}$, the potential energy U increases:

$$\boldsymbol{\tau} = \boldsymbol{\mu} \times \boldsymbol{B}, \qquad dU = \mu B \sin\theta \, d\theta, \qquad U(\theta) = -\mu B \cos\theta = -\boldsymbol{\mu} \cdot \boldsymbol{B}. \tag{5.16}$$

According to classical electromagnetism, an electric current i produces a magnetic moment proportional to the area subtended by the current. If this current is due to a particle with charge e and velocity v moving along a circumference of radius r, then

$$\mu_l = i\mathcal{A} = \frac{ev}{2\pi r}\pi r^2 = \frac{e}{2M}L = \frac{e}{|e|}\frac{g_l \mu_B}{\hbar}L. \tag{5.17}$$

Thus the magnetic moment due to the orbital motion is proportional to the orbital angular momentum. In vector and operator notation,

$$\hat{\boldsymbol{\mu}} = \frac{e}{|e|}\frac{g_l \mu_B}{\hbar}\hat{\boldsymbol{L}}, \tag{5.18}$$

where $\mu_B \equiv |e|\hbar/2M$ is called the Bohr magneton (Table 13.1) and $g_l = 1$ is the orbital gyromagnetic ratio.

Therefore, the presence of a magnetic field displaces the energy of a particle by an amount proportional to the component of the angular momentum along the magnetic field (Zeeman effect). Classically, this change is a continuous function of the orientation of the angular momentum but, according to quantum mechanics, the projections of the angular momentum are discretized (5.13):

$$\Delta E_{m_l} = g_l \mu_B B m_l \,. \tag{5.19}$$

Moreover, an orbital angular momentum should give rise to an odd number $(2l+1)$ of energy eigenstates.

For a uniform magnetic field, there is no net force acting on the magnetic dipole. However, if the field has a gradient in the z-direction, the net force is

$$F_z = \frac{\partial}{\partial z}(\boldsymbol{\mu} \cdot \boldsymbol{B}) = \mu_z \frac{\partial B}{\partial z} \,. \tag{5.20}$$

Figure 5.3 is a sketch of the experimental setup used by Stern and Gerlach [30]. Silver atoms are heated in an oven and escape through a hole. The beam is collimated and subsequently deflected by a nonuniform magnetic field perpendicular to its direction. Finally, a visible deposit is allowed to build up on a glass plate located far from the region of deflection.

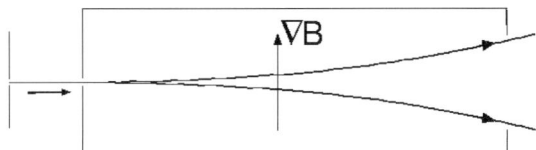

Fig. 5.3. Sketch of the Stern–Gerlach experimental arrangement

We may ignore the nuclear contributions to the magnetic moment on the grounds that the nuclear Bohr magneton is about 2000 times smaller. (It includes the proton mass in the denominator, instead of the electron mass.) Moreover, 46 of the 47 electrons form a spherically symmetric electron cloud with no net angular momentum (see Sect. 7.3). Therefore, the total spin of the Ag atom may be ascribed to the last electron.

The Stern–Gerlach result is reproduced in Fig. 5.4. Neither the classical continuous pattern nor the orbital quantum mechanical results displaying the separation into an odd number of terms was obtained: the beam was split into only two other beams, as would befit an angular momentum with $j = s = 1/2$.

5.2.2 Spin Formalism

Three years after the publication of the Stern–Gerlach experiment, George Uhlenbeck and Samuel Goudsmit proposed another quantum number in order

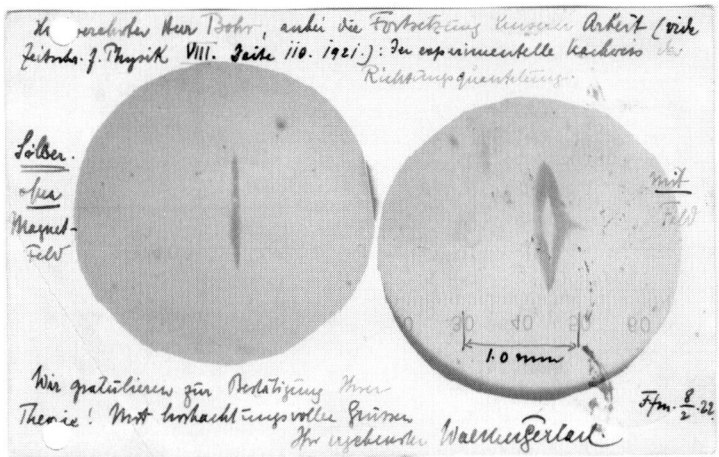

Fig. 5.4. Two figures contained in a letter to Bohr from Stern, communicating his experimental results. Stern explains that the magnetic field was too weak at the extremes of the beam. The figure to the left was obtained, for comparison, in the absence of magnetic field. (Reproduced with permission from Niels Bohr Archive, Copenhagen)

to specify the state of electrons (and of many other fundamental particles) [31]. It corresponds to a new physical entity, an intrinsic angular momentum called spin.

Since the spin is a pure quantum observable, only the matrix treatment formalism is possible. If $s = 1/2$, then the basis set of states is given by the two-component vectors (3.21) [32]

$$\varphi_{\uparrow z} \equiv \begin{pmatrix} 1 \\ 0 \end{pmatrix}, \qquad \varphi_{\downarrow z} \equiv \begin{pmatrix} 0 \\ 1 \end{pmatrix}. \tag{5.21}$$

The following representation of the spin operators reproduces (5.7) and (5.9) for $j = 1/2$:

$$\hat{S}_i = \frac{\hbar}{2}\sigma_i, \qquad i = x, y, z, \qquad \hat{S}^2 = \frac{\hbar^2 3}{4}\mathcal{I}, \tag{5.22}$$

where the σ_i are called the Pauli matrices:

$$\sigma_x = \begin{pmatrix} 0 & 1 \\ 1 & 0 \end{pmatrix}, \quad \sigma_y = \begin{pmatrix} 0 & -i \\ i & 0 \end{pmatrix}, \quad \sigma_z = \begin{pmatrix} 1 & 0 \\ 0 & -1 \end{pmatrix}, \quad \mathcal{I} = \begin{pmatrix} 1 & 0 \\ 0 & 1 \end{pmatrix}. \tag{5.23}$$

They all square to the unit matrix $\mathcal{I}$. The two-component states satisfy the eigenvalue equation

$$\begin{pmatrix} \langle\uparrow|Q|\uparrow\rangle & \langle\uparrow|Q|\downarrow\rangle \\ \langle\downarrow|Q|\uparrow\rangle & \langle\downarrow|Q|\downarrow\rangle \end{pmatrix} \begin{pmatrix} c_\uparrow \\ c_\downarrow \end{pmatrix} = q \begin{pmatrix} c_\uparrow \\ c_\downarrow \end{pmatrix}, \tag{5.24}$$

where the Pauli matrices (5.23) are used in the construction of the matrix $(\langle i|Q|j\rangle)$. The solution to this equation is obtained as in (3.23) and (3.24).

The spin has its own associated magnetic moment

$$\boldsymbol{\mu}_s = \frac{g_s \mu_\nu}{\hbar} \hat{\boldsymbol{S}} , \qquad (5.25)$$

with a gyromagnetic ratio of $g_s = 2.00$ for electrons, $g_s = 5.58$ for protons and $g_s = -3.82$ for neutrons. The constant μ_ν stands for minus the Bohr magneton μ_B in the case of electrons, or for the nuclear magneton $\mu_p = e_p \hbar / 2 M_p$ in the case of protons and neutrons (Table 13.1), where e_p and M_p are the proton charge and mass, respectively. The total magnetic moment operator is given by

$$\hat{\boldsymbol{\mu}} = \hat{\boldsymbol{\mu}}_s + \hat{\boldsymbol{\mu}}_l = \frac{\mu_\nu}{\hbar} \left(g_s \hat{\boldsymbol{S}} + g_l \hat{\boldsymbol{L}} \right) . \qquad (5.26)$$

Obviously, $g_l = 0$ for neutrons.

The eigenstates of the operator $\hat{S}_x$ have been obtained by means of a unitary transformation of the eigenvectors of $\hat{S}_z$ (3.26). This is a particular case of the more general transformation aligning the spin $s = 1/2$ operator with a direction of space labeled by the angles β, ϕ. The operator $\hat{S}_{\beta\phi}$ may be written as the scalar product of the spin vector $\hat{\boldsymbol{S}}$ times a unit vector along the chosen direction (see Problem 5 in Chap. 3):

$$\hat{S}_{\beta\phi} = \sin\beta\cos\phi \hat{S}_x + \sin\beta\sin\phi \hat{S}_y + \cos\beta \hat{S}_z$$
$$= \frac{\hbar}{2} \begin{pmatrix} \cos\beta & \sin\beta\exp(-i\phi) \\ \sin\beta\exp(i\phi) & -\cos\beta \end{pmatrix} . \qquad (5.27)$$

Upon diagonalization, we obtain the same two eigenvalues $\pm\hbar/2$. As explained in Sect. 3.1.4, this is a consequence of space isotropy. From this diagonalization, we also derive the state vectors

$$\varphi_{\uparrow\beta\phi} = \begin{pmatrix} \cos\frac{\beta}{2} \\ \sin\frac{\beta}{2}\exp(i\phi) \end{pmatrix}_z , \quad \varphi_{\downarrow\beta\phi} = \begin{pmatrix} -\sin\frac{\beta}{2}\exp(-i\phi) \\ \cos\frac{\beta}{2} \end{pmatrix}_z . \qquad (5.28)$$

The factor $1/2$ multiplying the angle β is characteristic of the effect of rotations on $s = 1/2$ objects. It may be verified through the value of the amplitudes (3.26) in the case of transformation from the z to the x eigenstates. In that case, $\beta = \pi/2, \phi = 0$.

An arbitrary linear combination of spin up and spin down states such as in (5.28) is called a qubit. The word 'qubit' is short for 'quantum bit', a concept used in quantum computation (Sect. 10.4).

5.3 Addition of Angular Momenta

Consider two angular momentum vector operators, $\hat{\boldsymbol{J}}_1$ and $\hat{\boldsymbol{J}}_2$. They are independent vectors, i.e., $[\hat{\boldsymbol{J}}_1, \hat{\boldsymbol{J}}_2] = 0$. Therefore, the product states are simultaneous eigenstates of the operators $\hat{J}_1^2, \hat{J}_{z1}, \hat{J}_2^2$ and $\hat{J}_{z2}$:

$$\varphi_{j_1 m_1 j_2 m_2} = \varphi_{j_1 m_1} \varphi_{j_2 m_2} \ . \tag{5.29}$$

These $(2j_1 + 1)(2j_2 + 1)$ eigenstates constitute a complete basis for states carrying the quantum numbers j_1, j_2. However, it may not be the most useful one. We may prefer a basis labeled by the quantum numbers associated with the total angular momentum $\hat{\boldsymbol{J}}$ (see Fig. 5.5):

$$\hat{\boldsymbol{J}} = \hat{\boldsymbol{J}}_1 + \hat{\boldsymbol{J}}_2 \ . \tag{5.30}$$

Since the components $\hat{J}_x, \hat{J}_y, \hat{J}_z$ also satisfy the commutation relations (5.4) and (5.6), there must also exist another basis set made up from eigenstates of the operators $\hat{J}^2$ and $\hat{J}_z$. Since the commutation relations

$$[\hat{J}^2, \hat{J}_1^2] = [\hat{J}^2, \hat{J}_2^2] = [\hat{J}^2, \hat{J}_z] = [\hat{J}_1^2, \hat{J}_z] = [\hat{J}_2^2, \hat{J}_z] = 0 \tag{5.31}$$

vanish, the new set of basis states may be labeled by the quantum numbers j_1, j_2, j, m. We apply a unitary transformation from the basis set (5.29) to the new basis (Sect. 3.1.3):

$$\varphi_{j_1 j_2 j m} \equiv \left[\varphi_{j_1} \varphi_{j_2}\right]_m^j = \sum_{m_1 + m_2 = m} c(j_1 m_1; j_2 m_2; j m) \varphi_{j_1 m_1 j_2 m_2} \ . \tag{5.32}$$

The quantum numbers j_1, j_2 are valid for both basis sets and they are not therefore summed up in (5.32).

For classical vectors, the modulus of the sum of two vectors lies between the sum of their moduli and the absolute value of their difference. Projections are simply added. Something similar takes place in quantum mechanics:

$$j_1 + j_2 \geq j \geq |j_1 - j_2| \ , \qquad m = m_1 + m_2 \ . \tag{5.33}$$

The quantum number j is an integer if both j_1, j_2 are integers or half-integers; j is a half-integer if only one of the constituents is an integer. The amplitudes $c(j_1 m_1; j_2 m_2; j m)$ are called Wigner or Clebsch–Gordan coefficients. They satisfy the symmetry relations

$$\begin{aligned} c(j_1 m_1; j_2 m_2; j m) &= (-1)^{j_1 + j_2 - j} c\big(j_1(-m_1); j_2(-m_2); j(-m)\big) \\ &= (-1)^{j_1 + j_2 - j} c(j_2 m_2; j_1 m_1; j m) \\ &= (-1)^{j_1 - m_1} \sqrt{\frac{2j+1}{2j_2+1}} c\big(j_1 m_1; j(-m); j_2(-m_2)\big) \ . \end{aligned} \tag{5.34}$$

The inverse transformation is

$$\varphi_{j_1 m_1} \varphi_{j_2 m_2} = \sum_{j=|j_1-j_2|}^{j=j_1+j_2} c(j_1 m_1; j_2 m_2; j m) \left[\varphi_{j_1} \varphi_{j_2}\right]_{m=m_1+m_2}^j \ . \tag{5.35}$$

The example of the summation of an angular momentum j_1 with the spin $j_2 = s_2 = 1/2$ is given in Sect. 5.6*.

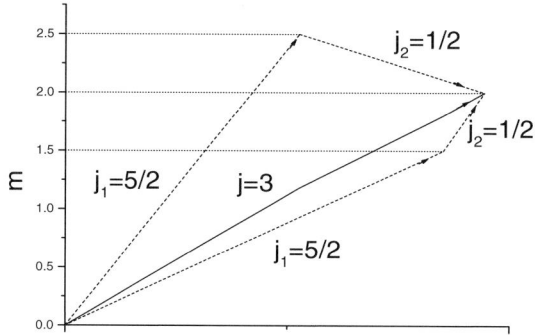

Fig. 5.5. Coupling of two vectors with $j_1 = 5/2$ and $j_2 = 1/2$ (*dotted lines*) may yield a vector with $j = 3$ (*continuous line*)

5.4* Details of Matrix Treatment

We define the operators
$$\hat{J}_\pm = \hat{J}_x \pm i\hat{J}_y \ . \tag{5.36}$$

They play a role similar to the creation and destruction operators a^+, a in the harmonic oscillator case (Sect. 3.2.1). Since the operator $\hat{J}_-$ is Hermitian conjugate to $\hat{J}_+$ (Sect. 2.7*),
$$\langle jm|J_+|jm'\rangle = \langle jm'|J_-|jm\rangle^* \ . \tag{5.37}$$

Applying the commutation relations (5.6), we obtain the relations
$$[\hat{J}_z, \hat{J}_+] = \hbar\hat{J}_+ \ , \tag{5.38}$$
$$[\hat{J}_+, \hat{J}_-] = 2\hbar\hat{J}_z \ . \tag{5.39}$$

The matrix elements of (5.38) read
$$\langle jm'|[J_z, J_+]|jm\rangle = \hbar(m' - m)\langle jm'|J_+|jm\rangle = \hbar\langle jm'|J_+|jm\rangle \ , \tag{5.40}$$

which implies that $\langle jm'|J_+|jm\rangle$ are only different from zero if $m' = m + 1$. Therefore the operator $\hat{J}_+$ raises the projection of the angular momentum by one unit of $\hbar$ (and $\hat{J}_-$ does the opposite).

The expectation value of (5.39) yields
$$\begin{aligned}\langle jm|[J_+, J_-]|jm\rangle &= \langle jm|J_+|j(m-1)\rangle\langle j(m-1)|J_-|jm\rangle \\ &\quad - \langle jm|J_-|j(m+1)\rangle\langle j(m+1)|J_+|jm\rangle \\ &= |\langle jm|J_+|j(m-1)\rangle|^2 - |\langle j(m+1)|J_+|jm\rangle|^2 \\ &= 2\hbar^2 m \ ,\end{aligned} \tag{5.41}$$

where (5.37) has been used. The solution to this first order difference equation in $|\langle j(m+1)|J_+|jm\rangle|^2$ is

$$|\langle j(m+1)|J_+|jm\rangle\rangle|^2 = \hbar^2[c - m(m+1)] \ . \tag{5.42}$$

Since the left-hand side is positive, only the values of m that make the right-hand side positive are allowed and the matrix element between the last allowed eigenstate $|\varphi_{jm_{\max}}\rangle$ and the first rejected eigenstate $|\varphi_{j(m_{\max}+1)}\rangle$ should therefore vanish. Here $m_{\max}$ is the positive root of the equation $c = m(m+1)$. The assignment of the quantum number $j = m_{\max}$ determines the value of the constant $c = j(j+1)$. Therefore,

$$\langle j(m+1)|J_+|jm\rangle = \hbar\sqrt{(j-m)(j+m+1)} \ , \tag{5.43}$$

where the positive value for the square root is chosen by convention. We verify the vanishing of the matrix elements connecting admitted and rejected states:

$$\langle j(j+1)|J_+|jj\rangle = \langle j(-j)|J_+|j(-j-1)\rangle = 0 \ . \tag{5.44}$$

Since m increases in steps of one unit between $-j$ and j [see (5.40)], the possible values of the quantum numbers j, m are those given in (5.8).

The matrix elements (5.9) corresponding to the operators $\hat{J}_x$ and $\hat{J}_y$ can be obtained from (5.37) and from (5.43). Addition of the squares of these matrices yields the (diagonal) matrix elements of $\hat{J}^2$ (5.7).

5.5* Details of the Treatment of Orbital Angular Momentum

Eigenvalue Equation for the Operator $\hat{L}_z$

The eigenvalue equation for the operator $\hat{L}_z$ is

$$-i\hbar\frac{\mathrm{d}\Psi(\phi)}{\mathrm{d}\phi} = l_z\Psi(\phi) \ . \tag{5.45}$$

The solution is proportional to $\exp(il_z\phi/\hbar)$. We may require $\Psi(\phi + 2\pi) = \Psi(\phi)$, which implies the existence of discrete values for the eigenvalue $l_z = \hbar m_l$ ($m_l = 0, \pm 1, \pm 2, \ldots$). Thus the orthonormal set of eigenfunctions of the operator $\hat{L}_z$ is given by

$$\varphi_{m_l}(\phi) = \frac{1}{\sqrt{2\pi}}\exp(im_l\phi) \ . \tag{5.46}$$

Eigenvalue Equation for the Operators $\hat{L}^2, \hat{L}_z$

We try a function of the form $\Psi(\theta, \phi) = P_{lm_l}(\theta)\exp(im_l\phi)$. It follows that, in the eigenvalue equation for the operator $\hat{L}^2$:

- we can make the replacement $\mathrm{d}^2/\mathrm{d}\phi^2 \to -m_l^2$,

- we may drop the exponential $\exp(im_l\phi)$ from both sides of the equation.

We obtain a differential equation depending on the single variable θ:

$$-\hbar^2\left(\frac{d^2}{d\theta^2} + \cot\theta\frac{d}{d\theta} - \frac{m_l^2}{\sin^2\theta}\right)P_{lm_l}(\theta) = \zeta P_{lm_l}(\theta). \tag{5.47}$$

The solutions to this equation for $m_l = 0$ may be expressed as polynomials $P_l(\cos\theta)$ of order l in $\cos\theta$, called Legendre polynomials ($l = 0, 1, 2, \ldots$). Each P_l gives rise to the $2l + 1$ associated Legendre functions $P_{lm_l}(\theta)$ with $|m_l| \le l$. All of them are eigenfunctions of the operator $\hat{L}^2$ with eigenvalue $\zeta = l(l + 1)\hbar^2$.

The simultaneous eigenfunctions of the operators $\hat{L}^2$ and $\hat{L}_z$ are called spherical harmonics:

$$Y_{lm_l}(\theta, \phi) = N_{lm_l} P_{lm_l}(\theta) \exp(im_l\phi), \tag{5.48}$$

where N_{lm_l} are constants chosen to satisfy the orthonormalization equation

$$\langle l'm_l'|lm_l\rangle = \int_0^\pi \sin\theta d\theta \int_0^{2\pi} d\phi Y_{l'm_l'}^* Y_{lm_l} = \delta_{ll'}\delta_{m_l m_l'}. \tag{5.49}$$

Here

$$Y_{lm_l}^* = (-1)^m Y_{l(-m_l)}. \tag{5.50}$$

The spherical harmonics corresponding to the lower values of l are given in Table 5.1. The l values (5.14) are traditionally replaced by symbolic letters in the literature (Table 5.2). This correspondence has only historical support.

Table 5.1. Spherical harmonics corresponding to the lowest values of l

$Y_{00} = \dfrac{1}{\sqrt{4\pi}}$	$Y_{1(\pm 1)} = \mp\sqrt{\dfrac{3}{8\pi}} \sin\theta \exp(\pm i\phi)$
$Y_{10} = \sqrt{\dfrac{3}{4\pi}} \cos\theta$	$Y_{2(\pm 1)} = \mp\sqrt{\dfrac{15}{32\pi}} \sin(2\theta) \exp(\pm i\phi)$
$Y_{20} = \sqrt{\dfrac{5}{16\pi}}(3\cos^2\theta - 1)$	$Y_{2(\pm 2)} = \sqrt{\dfrac{15}{32\pi}} \sin^2\theta \exp(\pm i2\phi)$

The coupling to angular momentum zero of two spherical harmonics depending on different orientations in space depends on the angle α_{12} subtended by the two orientations through the equation

$$\left[Y_l(\theta_1, \phi_1)Y_l(\theta_2\phi_2)\right]_0^0 = \frac{1}{\sqrt{2l+1}} \sum_{m_l=-l}^{m_l=l} (-1)^{l-m_l} Y_{lm_l}(\theta_1, \phi_1) Y_{l(-m_l)}(\theta_2, \phi_2)$$

$$= (-1)^l \frac{\sqrt{2l+1}}{4\pi} P_l(\cos\alpha_{12}). \tag{5.51}$$

Table 5.2. Equivalence between quantum number l and symbolic letters

l	Symbol
0	s
1	p
2	d
3	f
4	g

5.6* Coupling with Spin $s = 1/2$

The use of (5.32) is exemplified in the case where the second angular momentum is the spin $j_2 = s = 1/2$ (Fig. 5.5). Here the summation consists of two terms, corresponding to the two values of the spin projection $m_s = \pm 1/2$. According to (5.33), there are also two values for the total angular momentum $j = j_1 \pm 1/2$. However, if $j_1 = 0$, only the value $j = 1/2$ is allowed and there is a single term in (5.32):

$$\varphi_{(j_1=j+\frac{1}{2})sjm} = -\sqrt{\frac{j-m+1}{2j+2}}\varphi_{j_1(m-\frac{1}{2})_1}\begin{pmatrix}1\\0\end{pmatrix}_2 \tag{5.52}$$

$$+\sqrt{\frac{j+m+1}{2j+2}}\varphi_{j_1(m+\frac{1}{2})_1}\begin{pmatrix}0\\1\end{pmatrix}_2,$$

$$\varphi_{(j_1=j-\frac{1}{2})sjm} = \sqrt{\frac{j+m}{2j}}\varphi_{j_1(m-\frac{1}{2})_1}\begin{pmatrix}1\\0\end{pmatrix}_2 + \sqrt{\frac{j-m}{2j}}\varphi_{j_1(m+\frac{1}{2})_1}\begin{pmatrix}0\\1\end{pmatrix}_2.$$

A particular application of this example is the coupling of orbital motion with the spin of an electron (Sect. 6.2). In this case the eigenstates $\varphi_{j_1 m_1}$ are the spherical harmonics $Y_{l_1 m_{l_1}}$ (5.48). However, (5.52) is valid whatever the nature of the angular momentum j_1 may be.

Problems

Problem 1. A plastic disk rotates with angular velocity 100 rad/s. Estimate, in units of $\hbar$, the order of magnitude of the angular momentum.

Problem 2.

1. Construct the matrix for the operator $\hat{L}_x$ (5.12) in the basis of spherical harmonics Y_{1m_l} (Table 5.1).
2. Diagonalize the matrix and compare its eigenvalues with those of the operator $\hat{L}_z$.

5 Angular Momentum

Problem 3. Verify that the product of the uncertainties $\Delta_{J_x}\Delta_{J_y}$ satisfies the inequality (2.27).

Problem 4. Calculate $\hat{\boldsymbol{J}}\times\hat{\boldsymbol{J}}$.

Problem 5. Consider the following matrix elements between spherical harmonic states:

$$\langle 00|Y_{20}|00\rangle, \quad \langle 10|Y_{20}|10\rangle, \quad \langle 11|Y_{21}|21\rangle, \quad \langle 00|Y_{11}|11\rangle, \quad \langle 00|Y_{11}|1(-1)\rangle,$$

$$\langle 00|\Pi|00\rangle, \quad \langle 11|\Pi|11\rangle, \quad \langle 00|\Pi|10\rangle.$$

1. Find out which of the above matrix elements vanishes due to conservation of orbital angular momentum and/or parity.
2. Calculate those that remain.

Problem 6. Calculate $[\hat{S}_x^2, \hat{S}_z]$ for spin $s = 1/2$ particles.

Problem 7.

1. Construct the eigenstates of $\hat{S}_x$ and $\hat{S}_y$ using the eigenstates of $\hat{S}_z$ as basis states.
2. If the spin S_x is measured when the particle is in an eigenstate of the operator $\hat{S}_y$, what are the possible results and their probabilities?
3. Construct the matrix corresponding to $\hat{S}_x$ using the eigenstates of $\hat{S}_y$ obtained in the first part as basis states.
4. Express the eigenstates $\varphi^{(s_x)}$ using the eigenstates $\varphi^{(s_y)}$ as basis states.

Problem 8. A particle is in the spin state $\begin{pmatrix} a \\ b \end{pmatrix}$, with a, b real. Calculate the probability of obtaining the eigenvalue $\hbar/2$ if:

1. S_x is measured,
2. S_y is measured,
3. S_z is measured.

Problem 9. A particle is in the spin state $\Psi = \begin{pmatrix} \cos(\theta/2) \\ \sin(\theta/2) \end{pmatrix}$.

1. What are the values of S_z that would appear as a result of a measurement of this observable? What are the associated probabilities?
2. What is the mean value of S_z in this state?

Problem 10.

1. Construct the possible states with $m = 1/2$ that are obtained by coupling an orbital angular momentum $l = 2$ with a spin $s = 1/2$.
2. Verify the orthonormality of the coupled states.
3. Construct the wave vector corresponding to the state with $j = m = l + 1/2$. What is the probability that the spin points up?

Problem 11. Write the two-spin state vectors with $s = \frac{1}{2}$ that have a definite total angular momentum.

Problem 12. Apply the closure property to the transformations (5.32) and (5.35), as in (3.16).

6 Three-Dimensional Hamiltonian Problems

6.1 Central Potentials

In the present chapter, we broaden the quantum mechanical treatment of the problem of a single particle moving in three-dimensional space so that it includes the treatment of the Hamiltonian.

The solution to a given problem can usually be simplified by exploiting the associated symmetries. We have already shown that this is the case by applying invariance under the inversion operation (see the bound problems of Sects. 4.2, 4.4 and 4.6*). Since problems involving a central potential $V(\mathbf{r}) = V(r)$ have spherical symmetry, we shall make use of this symmetry. For this purpose, we write the kinetic energy Laplacian in spherical coordinates (5.11). The total Hamiltonian reads

$$\hat{H} = \frac{1}{2M}\left(\hat{p}_x^2 + \hat{p}_y^2 + \hat{p}_z^2\right) + V(r)$$
$$= \frac{\hbar^2}{2M}\left(-\frac{\partial^2}{\partial r^2} - \frac{2}{r}\frac{\partial}{\partial r}\right) + \frac{\hat{L}^2}{2Mr^2} + V(r), \quad (6.1)$$

where the operator $\hat{L}^2$ is the square of the orbital angular momentum (5.12). Since the Hamiltonian (6.1) commutes with operators $\hat{L}^2$ and $\hat{L}_z$, there is a basis set of eigenfunctions for the three operators. The eigenvalue equation (4.10) is solved by factorizing the wave function into radial and angular terms, the latter being represented by the spherical harmonics (5.48):

$$\Psi(r,\theta,\phi) = R_{n_r l}(r) Y_{lm_l}(\theta,\phi). \quad (6.2)$$

In such a case, the operator $\hat{L}^2$ in the kinetic energy can be replaced by its eigenvalue $\hbar^2 l(l+1)$ and moreover, the spherical harmonics cancel on both sides of the Schrödinger equation.[1] One is left with a differential equation depending on a single variable: the radius. Thus,

$$\left\{\frac{\hbar^2}{2M}\left[-\frac{d^2}{dr^2} - \frac{2}{r}\frac{d}{dr} + \frac{l(l+1)}{r^2}\right] + V(r)\right\} R_{n_r l}(r) = E_{n_r l} R_{n_r l}(r), \quad (6.3)$$

---

[1] This is another application of the separation of variables method for solving partial differential equations.

where the new quantum number n_r distinguishes between states with the same value of l. Since the magnetic quantum number m_l does not appear in this equation, the eigenvalues $E_{n_r l}$ are also independent of it. In consequence, the eigenenergies of a central potential are necessarily degenerate, with degeneracy equal to $2l+1$ (5.14). This result is to be expected since the quantum number m_l depends on the orientation of the coordinate axis. That is to say, the central potential has spherical symmetry and the resulting energies (which are physical quantities) should not depend on the orientation of the coordinate axis (which is an artifact of the calculation).

6.1.1 Coulomb and Harmonic Oscillator Potentials

In this section we discuss the solutions to the eigenvalue equation for two central potentials: the Coulomb potential $-Ze^2/4\pi\epsilon_0 r$ and the three-dimensional harmonic oscillator potential $M\omega^2 r^2/2$.

It is always useful to begin by estimating the orders of magnitude of the quantities involved. For the linear harmonic oscillator, this has already been done in (3.33). These orders of magnitude remain valid for the three-dimensional case, since the harmonic Hamiltonian is separable into three Cartesian coordinates, and the estimate (3.33) holds for each coordinate. For the Coulomb potential, we may again use the Heisenberg uncertainty relations

$$\boldsymbol{p}^2 \approx 3\Delta p_x^2 \geq \frac{3\hbar^2}{4}\frac{1}{\Delta x^2} \approx \frac{9\hbar^2}{4}\frac{1}{r^2}. \qquad (6.4)$$

Therefore, the radius r_m is obtained by minimizing the lower bound energy

$$E \geq \frac{9\hbar^2}{8Mr^2} - \frac{Ze^2}{4\pi\epsilon_0 r}, \qquad (6.5)$$

which yields

$$r_\mathrm{m} = \frac{9}{4Z}a_0, \qquad E \geq \frac{16Z^2}{9}E_\mathrm{H}, \qquad (6.6)$$

where the Bohr radius a_0 and the ground state energy of the hydrogen atom E_H are given in Tables 6.1 and 13.1.

The solutions of the Schrödinger equation for the Coulomb and the harmonic oscillator potentials are shown in Table 6.1. The corresponding details are outlined in Sect. 6.4*. The following comments stem from the comparison between the solutions for these two potentials:

- In both cases the radial factor $R_{n_r l}(r)$ may be expressed as a product of an exponential decay, a power of u, u^l, and a polynomial of degree n_r (Coulomb) or $2n_r$ (harmonic oscillator).
- The radial factor u^l decreases the radial density $|R_{n_r l}|^2 r^2$ for small values of u and increases it for large values. It is a manifestation of centrifugal effects due to rotation of the particle.

Table 6.1. Solutions to the Coulomb and harmonic oscillator problems

Problem	Coulomb	Harmonic oscillator
Characteristic length	$a_0 = 4\pi\epsilon_0 \hbar^2/Me^2$	$x_c = \sqrt{\hbar/M\omega}$
Wave function	$R_{n_r l}(u)Y_{lm_l}(\theta,\phi)$ $u = Zr/a_0$	$R_{n_r l}(u)Y_{lm_l}(\theta,\phi)$ $u = r/x_c$
Radial quantum numbers	$n_r = 0, 1, \ldots$	$n_r = 0, 1, \ldots$
Principal quantum numbers	$n = n_r + l + 1 = 1, 2, \ldots$	$N = 2n_r + l = 0, 1, \ldots$
Energies	$Z^2 E_H/n^2$ $E_H = -e^2/8\pi\epsilon_0 a_0$	$\hbar\omega(N + 3/2)$
Degeneracy	n^2	$(N+1)(N+2)/2$

- Both potentials display a higher degree of degeneracy than is required by spherical invariance.
- All degenerate states in the harmonic oscillator potential have the same value of $(-1)^l = (-1)^N$, where N is the principal quantum number (Table 6.1), and thus have the same parity. This is not true for the Coulomb potential, where states with even and odd values of l may be degenerate [see the last equation of (5.13)].
- The energies of the Coulomb potential are represented in Fig. 6.1, while those of the harmonic potential have the same pattern as in Fig. 3.2. The eigenvalues of the former display an accumulation point at $E_\infty = 0$. They are equidistant in the harmonic oscillator case.

We have verified the commonly made statement that the Schrödinger equation is exactly soluble for the two central potentials treated in this section. In fact the two Schrödinger equations are related by a simple change of independent variable $r \to r^2$, if the energy and the strength of the potential are swapped and the orbital angular momentum is rescaled (Problem 8) [34]. Thus, the Schrödinger equations corresponding to the Coulomb and three-dimensional harmonic oscillator potential constitute only one soluble quantum mechanical central problem, not two.

The harmonic oscillator potential is also separable in Cartesian coordinates. As an exercise, derive the degeneracies using the Cartesian solution and check the results against those appearing in the last column of Table 6.1.

6.2 Spin–Orbit Interaction

One may incorporate the spin degree of freedom into the present treatment. The degeneracies displayed in Table 6.1 are thus doubled.

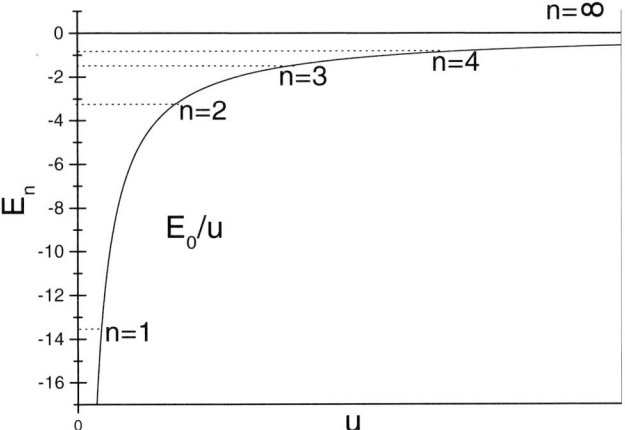

Fig. 6.1. Coulomb potential and its eigenvalues. The dimensionless variable $u = r/a_0$ has been used

According to the results of Sect. 5.3, there are two complete sets of wave functions that may take care of the spin $s = 1/2$:

$$\varphi_{nlm_l s m_s} = R_{n_r l} Y_{lm_l} \varphi_{sm_s} \,, \tag{6.7}$$

$$\varphi_{nlsjm} = R_{n_r l} \sum_{m_l + m_s = m} c(lm_l; sm_s; jm) Y_{lm_l} \varphi_{sm_s} \,, \tag{6.8}$$

where the Clebsch–Gordan or Wigner coefficients are given in Sect. 5.6*. As mentioned in Sect. 5.3, the first set is labeled with the quantum numbers $lm_l s m_s$ specifying the modulus and the z-projection of the orbital angular momentum and the spin. In the second set, the moduli of the orbital angular momentum and the spin remain as good quantum numbers, to be accompanied by jm, associated with the modulus and z-projection of the total angular momentum $\hat{\boldsymbol{J}} = \hat{\boldsymbol{L}} + \hat{\boldsymbol{S}}$.

The Coulomb interaction is the strongest force acting inside an atom, and yields adequate results for many purposes. However, the experimental spectrum displays small shifts in energy associated with values of j. Another (weaker) force that is present in the atom is provided by the interaction between the magnetic moment of the spin and the magnetic field produced by the orbital motion of the electron:[2]

$$\hat{V}_{\text{so}} = v_{\text{so}} \hat{\boldsymbol{S}} \cdot \hat{\boldsymbol{L}} \,, \tag{6.9}$$

---

[2] Criteria which are frequently used to construct interactions involving quantum variables are: (i) simplicity and (ii) invariance under rotations, parity and time-reversal transformations. The interaction (6.9) satisfies all these criteria. Moreover, it may also be obtained in the non-relativistic limit of the Dirac equation.

where we have approximated the radial factor by a constant v_{so}. There are additional terms, the hyperfine interactions, arising from the interaction between the nuclear and the electron spins. Although they are even smaller, they produce a splitting of the ground state of the hydrogen atom with astrophysical importance [which the interaction (6.9) does not].

Suppose we sit on the electron. We see the charged nucleus orbiting around us. The current associated with this moving charged nucleus produces a magnetic field at the location of the electron. The $\hat{\boldsymbol{S}}\cdot\hat{\boldsymbol{L}}$ term can be interpreted as the interaction between the spin magnetic moment of the electron and this magnetic field.

The radial term $R_{n_r l}$ has been dropped in the present section, since the spin–orbit interaction (6.9) does not affect the radial part of the wave function.

The spin–orbit interaction satisfies the commutation relations

$$[\hat{V}_{\text{so}}, \hat{L}^2] = [\hat{V}_{\text{so}}, \hat{S}^2] = [\hat{V}_{\text{so}}, \hat{J}^2] = [\hat{V}_{\text{so}}, \hat{J}_z] = 0 , \qquad (6.10)$$

while $[\hat{V}_{\text{so}}, \hat{L}_z] \neq 0$, $[\hat{V}_{\text{so}}, \hat{S}_z] \neq 0$. Bearing in mind this property, different procedures – already developed in these notes – may be applied in order to incorporate the interaction (6.9).

1. The interaction is not diagonal within the set of eigenstates of the projections of the angular momenta (6.7). Since $\hat{V}_{\text{so}}$ commutes with $\hat{J}_z$, the spin–orbit interaction conserves the total projection $m = m_l + m_s$ and thus gives rise to matrices of order 2 which may be diagonalized according to Sect. 3.1.4.
2. The spin–orbit interaction is diagonal within the set of eigenstates (6.8). This constitutes a significant advantage. The diagonal matrix elements are the eigenvalues, which may be obtained through calculation.
3. Observing that

$$\hat{\boldsymbol{L}}\cdot\hat{\boldsymbol{S}} = \frac{1}{2}\left(\hat{J}^2 - \hat{L}^2 - \hat{S}^2\right) , \qquad (6.11)$$

we obtain

$$\langle lsjm|\boldsymbol{S}\cdot\boldsymbol{L}|lsjm\rangle = \frac{\hbar^2}{2}\left[j(j+1) - l(l+1) - \frac{3}{4}\right] . \qquad (6.12)$$

Due to the spin–orbit interaction, the two states with $j_\pm = l \pm 1/2$ become displaced by an amount proportional to the values appearing on the right-hand side of (6.12).

6.3 Some Elements of Scattering Theory

6.3.1 Boundary Conditions

We consider an incident particle scattered by a central, finite-sized potential. The asymptotic boundary condition for this problem requires the asymptotic wave function to be expressed as a superposition of an incident plane wave along the z-axis, and an outgoing spherical wave (Fig. 6.2):

$$\lim_{r\to\infty} \Psi(r,\theta) = A\left[\exp(ikz) + \frac{\exp(ikr)}{r}f_k(\theta)\right], \quad (6.13)$$

where $k = \sqrt{2ME}/\hbar$ is the wave number (4.30) and $f_k(\theta)$ is the amplitude of the scattered wave in the polar direction θ. The spherical wave carries a factor $1/r$, since $|\Psi(r)|^2$ must be proportional to $1/r^2$ in order to conserve probability (see Problem 11). The azimuth angle ϕ does not appear, because the problem displays axial symmetry. Expression (6.13) constitutes a generalization of the boundary conditions discussed at the beginning of Sect. 4.5 to the three-dimensional case.

6.3.2 Expansion in Partial Waves

As in the case of a three-dimensional harmonic oscillator, the free particle problem admits solutions in both Cartesian and polar coordinates. The solutions to the Hamiltonian (6.1) with $V(r) = 0$ are, in spherical coordinates,

$$\varphi^{(1)}_{lm_l}(r,\theta,\phi) = j_l(kr)Y_{lm_l}(\theta,\phi), \quad \varphi^{(2)}_{lm_l}(r,\theta,\phi) = n_l(kr)Y_{lm_l}(\theta,\phi), \quad (6.14)$$

where j_l and n_l are Bessel and Neumann functions, respectively (see Sect. 6.5*). The eigenstates (6.14) constitute a complete set. Our immediate task is to construct the most general linear combination which asymptotically yields (6.13). We first note that the function $\exp(ikz)$ may be expanded as

$$\exp(ikz) = \sqrt{4\pi}\sum_{l=0}^{l=\infty} i^l(2l+1)^{1/2} j_l Y_{l0}. \quad (6.15)$$

Secondly, the second term on the right-hand side of (6.13) can be written in terms of the Hankel function of the first kind, which behaves asymptotically as an outgoing spherical wave (6.34). Therefore, the most general and acceptable linear combination is

$$\Psi(r,\theta) = A\sum_{l=0}^{l=\infty}\left[\sqrt{4\pi}i^l(2l+1)^{1/2}j_l + c_l h_l^{(+)}(kr)\right]Y_{l0}$$

$$= A\sqrt{\pi}\sum_{l=0}^{l=\infty} i^l(2l+1)^{1/2} a_l \left(j_l\cos\delta_l - n_l\sin\delta_l\right)Y_{l0}. \quad (6.16)$$

Here c_l, a_l are complex amplitudes, which may be expressed in terms of δ_l, the (real) phase shift of the l-partial wave:

$$c_l = \sqrt{\pi}i^l(2l+1)^{1/2}(a_l^2-1), \quad a_l = \exp(i\delta_l). \quad (6.17)$$

We notice that $f_k(\theta)$ is provided by the second term in the first line of (6.16). Replacing the Hankel function by its asymptotic representation one gets

$$f_k(\theta) = -\mathrm{i}\frac{\sqrt{\pi}}{k}\sum_{l=0}^{l=\infty}(2l+1)^{1/2}\left[\exp(\mathrm{i}2\delta_l) - 1\right]Y_{l0}\ . \tag{6.18}$$

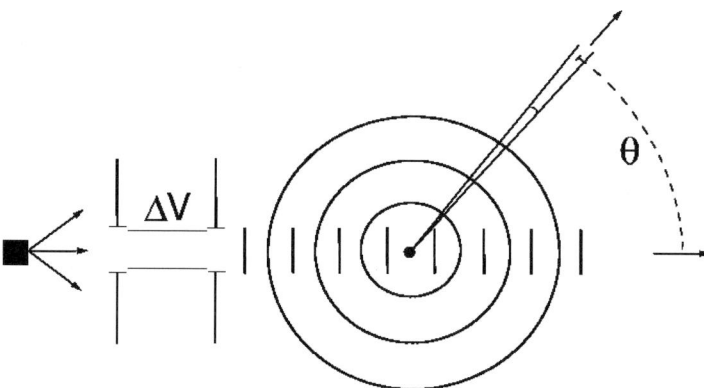

Fig. 6.2. Schematic representation of a scattering experiment. After being produced in a source, the projectile is collimated, accelerated, and collimated again. It collides with the target in the form of a plane wave. It is subsequently scattered as a spherical wave, within a solid angle that makes an angle θ with the direction of incidence

6.3.3 Cross-Sections

According to (6.13), the ratio between the scattered flux in the direction θ and the incident flux along the polar axis, is given by $|f(\theta)|^2/r^2$. The differential cross-section is defined as the number of particles that emerge per unit incident flux, per unit solid angle, and per unit time:

$$\sigma(\theta) = |f(\theta)|^2 = \frac{4\pi}{k^2}\left|\sum_{l=0}^{l=\infty}(2l+1)^{1/2}\left[\exp(\mathrm{i}2\delta_l) - 1\right]Y_{l0}\right|^2\ . \tag{6.19}$$

The total cross-section is the integral over the whole solid angle

$$\sigma = 2\pi\int_0^\pi \sigma(\theta)\sin\theta\mathrm{d}\theta = \frac{4\pi}{k^2}\sum_{l=0}^{l=\infty}(2l+1)\sin^2\delta_l\ . \tag{6.20}$$

The values of δ_l are determined by applying continuity equations at the border $r = a$ of the central potential. In the case of scattering by a rigid sphere of radius a, the phase shifts are given by (6.16) and (6.32):

$$\tan \delta_l = \frac{j_l(ka)}{n_l(ka)}, \qquad \lim_{ka \to 0} \tan \delta_l = \frac{(ka)^{2l+1}}{(2l+1)\left[(2l-1)!!\right]^2}. \qquad (6.21)$$

If $ka = 0$, all the partial wave contributions vanish except for $l = 0$, due to the k^2 appearing in the denominator of the cross-sections (6.19) and (6.20). We obtain

$$\sigma(\theta) = a^2, \qquad \sigma = 4\pi a^2. \qquad (6.22)$$

The scattering is spherically symmetric and the total cross-section is four times the area seen by classical particles in a head-on collision. This quantum result also appears in optics, and is characteristic of long-wavelength scattering. The fact that σ is the total surface area of the sphere is interpreted by saying that the waves 'feel' all this area.

Some features of scattering theory deserve to be stressed:

- The classical distance of closest approach to the z-axis of a particle with orbital angular momentum $\hbar l$ and energy E is l/k. Therefore, a classical particle is not scattered if $l > ka$. A similar feature appears in quantum mechanics, since the first and largest maximum of $j_l(kr)$ lies approximately at $r = l/k$. Thus, for $l > ka$, the maximum occurs where the potential vanishes: the largest value of l to be included is of order ka.
- The calculation of the probability current (4.17) with wave function (6.13) should yield interference terms in the whole space. They would be non-physical consequences of assuming an infinite plane wave for the incident beam. In practice, the beam is collimated and, as a consequence, the incident plane wave and the scattered wave are well separated, except in the forward direction (Fig. 6.2). On the other hand, in most experimental arrangements, the opening of the collimator is sufficiently large to ensure that there are no measurable effects of the uncertainty principle due to collimation (see Problem 12 of Chap. 2).
- Interference in the forward direction between the incident plane wave and the scattered wave gives rise to the important relation

$$\sigma = \frac{4\pi}{k} \mathrm{Im}\left[f_k(0)\right], \qquad (6.23)$$

by comparing (6.18) and (6.20). The attenuation of the transmitted beam measured by $\mathrm{Im}[f_k(0)]$ is proportional to the total cross-section σ. The validity of (6.23) (optical theorem) is very general, and is not restricted to scattering theory.
- The previous description of a scattering experiment is made in the center-of-mass coordinate system. We must therefore use the projectile–target reduced mass and the energy for the relative motion in order to determine the value of k. Moreover, there is a geometrical transformation between the scattering angles θ and θ_{lab} because the two systems of reference move relative to each other with the velocity of the center of mass.

6.4* Solutions to the Coulomb and Oscillator Potentials

The hydrogen atom constitutes a two-body problem which can be transformed to a one-body form by changing to the center of mass frame. As a consequence, the reduced mass for relative motion should be used [as will be done in (8.23)]. However, for the sake of simplicity, we ignore the motion of the nucleus here, since it is much heavier than the electron.

It is always helpful to work with dimensionless variables, as in (4.21). In the case of the hydrogen atom, the natural length is the Bohr radius (Table 13.1). Thus, one may use $u = Zr/a_0$. The solution to the radial equation (6.3) takes the form

$$R_{n_r l}(r) = N_{nl}(Z/na_0)^{3/2} \exp(-u/n) u^l L_{n_r}(u) , \qquad (6.24)$$

where $L_{n_r}(u)$ are polynomials of degree $n_r = 0, 1, 2, \ldots$ called Laguerre polynomials. The N_{nl} are normalization constants such that

$$\int_0^\infty r^2 R_{n_r l} R_{n'_r l} \mathrm{d}r = \delta_{n_r n'_r} . \qquad (6.25)$$

The energy $Z^2|E_\mathrm{H}|/n^2$ is the ionization or binding energy, i.e., the amount of energy that must be given to the Z atom in order to separate an electron in the n state. Figure 6.3 represents the probability density as a function of the radial coordinate for the states included in Table 6.2. The expression for the probability density includes a factor r^2 associated with the volume element (5.11).

Figure 6.4 combines the angular distribution associated with the spherical harmonics of Fig. 5.2 with the radial densities appearing in Fig. 6.3.

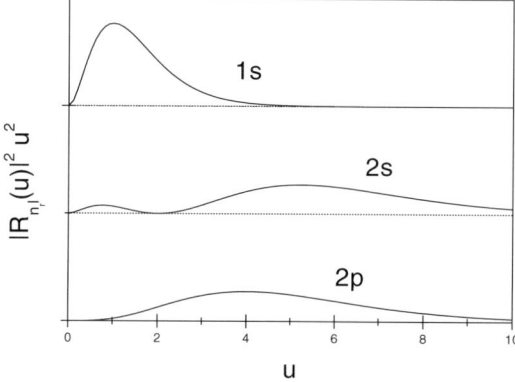

Fig. 6.3. Radial probability densities of the Coulomb potential

The Bohr radius a_0 may be compared with the expectation value of the coordinate r in the ground state of the hydrogen atom. According to Table 6.2, one gets

$$\langle 100|r|100\rangle = \int_0^\infty \int_0^\pi \int_0^{2\pi} r^3 \left|\varphi_{100}\right|^2 \mathrm{d}r \sin\theta \mathrm{d}\theta \mathrm{d}\phi$$
$$= \frac{4}{a_0^3} \int_0^\infty r^3 \exp(-2r/a_0)\mathrm{d}r = \frac{3}{2}a_0 \, . \tag{6.26}$$

There are also positive energy, unbound solutions to the Coulomb problem. They are used in the analysis of scattering experiments between charged particles.

In the harmonic oscillator, the dimensionless length is given by the ratio $u = r/x_c$, as in (4.20). The radial eigenfunctions are

Table 6.2. Radial dependence of the lowest solutions for the Coulomb potential and the three-dimensional harmonic oscillator

Coulomb potential				
n	n_r	l	N_{nl}	$L_{n_r}(u)$
1	0	0	2	1
2	1	0	2	$1 - \frac{1}{2}u$
2	0	1	$1/\sqrt{3}$	1
3	2	0	2	$1 - \frac{2}{3}u + \frac{2}{27}u^2$
3	1	1	$4\sqrt{2}/9$	$1 - \frac{1}{6}u$
3	0	2	$4/27\sqrt{10}$	1

Three-dimensional harmonic oscillator				
N	n_r	l	N_{Nl}	$F\left(-n_r, l+\frac{3}{2}, u^2\right)$
0	0	0	2	1
1	0	1	$2\sqrt{2/3}$	1
2	1	0	$\sqrt{6}$	$1 - \frac{2}{3}u^2$
2	0	2	$4/\sqrt{15}$	1
3	1	1	$2\sqrt{5/3}$	$1 - \frac{2}{5}u^2$
3	0	3	$4\sqrt{2/105}$	1

6.4* Solutions to the Coulomb and Oscillator Potentials 89

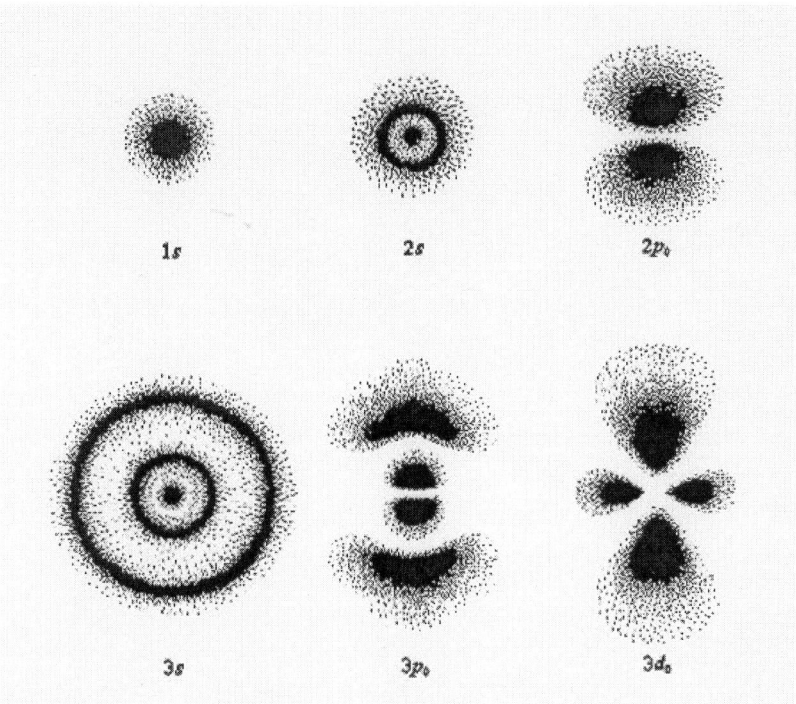

Fig. 6.4. Probability density plots of some hydrogen atomic orbitals. The density of the dots represents the probability of finding the electron in that region [33]. (Reproduced with permission from University Science Books)

$$R_{n_r l} = N_{Nl} \frac{1}{\pi^{1/4} x_c^{3/2}} \exp(-u^2/2) u^l F\left(-n_r, l + \frac{3}{2}, u^2\right) . \tag{6.27}$$

The confluent hypergeometric function $F(-n_r, l + 3/2, u^2)$ is a polynomial of the order n_r in u^2 ($n_r = 0, 1, 2, \ldots$). Some radial probability densities are displayed in Fig. 6.5. The N_{Nl} are normalization constants such that (6.25) also holds true in this case. The energy eigenvalues are given by

$$E = \hbar\omega \left(N + \frac{3}{2}\right) . \tag{6.28}$$

Using procedures similar to those applied for the linear harmonic oscillator, we may calculate the expectation values of the square of the radius and of the momentum. We thus verify the virial theorem (3.51) once again:

$$\langle Nlm_l|r^2|Nlm_l\rangle / x_c^2 = \langle Nlm_l|p^2|Nlm_l\rangle x_c^2/\hbar^2 = N + \frac{3}{2} . \tag{6.29}$$

The lowest energy solutions are given on the right-hand side of Table 6.2.

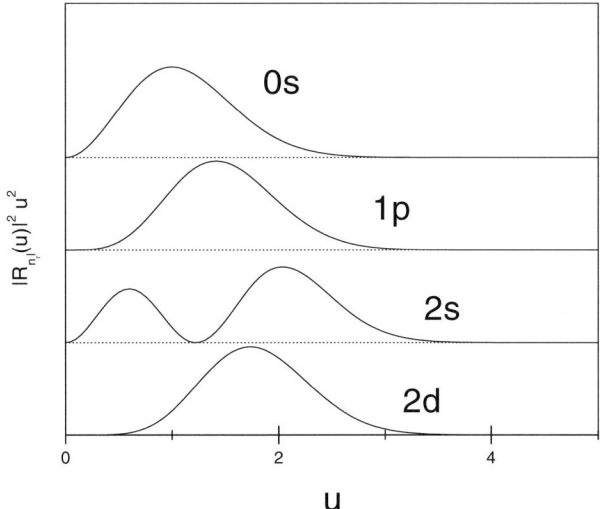

Fig. 6.5. Radial probability densities of the harmonic oscillator potential

Useful definite integrals are

$$\int_0^\infty u^n \exp(-au)\,du = \frac{n!}{a^{n+1}}, \qquad \int_0^\infty u^{2n} \exp(-u^2)\,du = \frac{(2n-1)!!\sqrt{\pi}}{2^{n+1}},$$

$$\int_0^\infty u^{2n+1} \exp(-u^2)\,du = \frac{n!}{2}.$$

6.5* Some Properties of Spherical Bessel Functions

The spherical Bessel functions $j_l(kr)$ [and the Neumann $n_l(kr)$] satisfy the differential equation

$$-\frac{\hbar^2}{2M}\left[\frac{d^2}{dr^2} + \frac{2}{r}\frac{d}{dr} - \frac{l(l+1)}{r^2}\right] j_l(kr) = \frac{\hbar^2 k^2}{2M} j_l(kr). \qquad (6.30)$$

Their asymptotic properties for large arguments are

$$\lim_{\rho \to \infty} j_l(\rho) = \frac{1}{\rho} \sin\left(\rho - \frac{1}{2}l\pi\right), \qquad \lim_{\rho \to \infty} n_l(\rho) = -\frac{1}{\rho} \cos\left(\rho - \frac{1}{2}l\pi\right),$$
$$(6.31)$$

while for small arguments they are

$$\lim_{\rho \to 0} j_l(\rho) = \frac{\rho^l}{(2l+1)!!}, \qquad \lim_{\rho \to 0} n_l(\rho) = -\frac{(2l-1)!!}{\rho^{l+1}}. \qquad (6.32)$$

The spherical Hankel functions are defined by

$$h_l^{(+)}(\rho) = j_l(\rho) + in_l(\rho) , \qquad h_l^{(-)}(\rho) = j_l(\rho) - in_l(\rho) . \tag{6.33}$$

Due to (6.31), these have the asymptotic expressions

$$\lim_{\rho\to\infty} h_l^{(+)}(\rho) = \frac{(-i)^{l+1}}{\rho} \exp(i\rho) , \qquad \lim_{\rho\to\infty} h_l^{(-)}(\rho) = \frac{(i)^{l+1}}{\rho} \exp(-i\rho) . \tag{6.34}$$

The first three j_ls and n_ls are given in Table 6.3.

Table 6.3. Lowest spherical Bessel functions

l	j_l	n_l
0	$\dfrac{1}{\rho}\sin\rho$	$-\dfrac{1}{\rho}\cos\rho$
1	$\dfrac{1}{\rho^2}\sin\rho - \dfrac{1}{\rho}\cos\rho$	$-\dfrac{1}{\rho^2}\cos\rho - \dfrac{1}{\rho}\sin\rho$
2	$\left(\dfrac{3}{\rho^3}-\dfrac{1}{\rho}\right)\sin\rho - \dfrac{3}{\rho^2}\cos\rho$	$-\left(\dfrac{3}{\rho^3}-\dfrac{1}{\rho}\right)\cos\rho - \dfrac{3}{\rho^2}\sin\rho$

Problems

Problem 1. Calculate the difference in the excitation energy of $n=2$ states between hydrogen and deuterium atoms. Hint: use the reduced mass instead of the electron mass.

Problem 2.

1. Assign the quantum numbers nlj to the eigenstates of the Coulomb problem with $n \leq 3$.
2. Do the same for the three-dimensional harmonic oscillator with $N \leq 3$.

Problem 3.

1. Obtain the degeneracy of a harmonic oscillator shell N, including the spin.
2. Obtain the average value $\langle|L^2|\rangle_N$ of the operator $\hat{L}^2$ in the same shell.
3. Calculate the eigenvalues of a harmonic oscillator potential plus the interaction [see (7.16)]

$$-\frac{1}{16}\frac{\omega}{\hbar}\left(\hat{L}^2 - \langle|L^2|\rangle_N\right) - \frac{1}{4}\frac{\omega}{\hbar}\hat{L}\cdot\hat{S} ,$$

for $N = 0, 1, 2, 3$.

4. Give the quantum numbers of the states with minimum energy for a given shell N.

Problem 4.

1. Find the energy and the wave function for a particle moving in an infinite spherical well of radius a with $l = 0$. Hint: replace $\Psi(r) \to f(r)/r$.
2. Solve the same problem using the Bessel functions given in Sect. 6.5*.

Problem 5.

1. Find the values of r at which the probability density is at a maximum, assuming the $n = 2$ states of a hydrogen atom.
2. Calculate the mean value of the radius for the same states.

Problem 6. Solve the harmonic oscillator problem in Cartesian coordinates. Calculate the degeneracies and compare them with those listed in Table 6.1.

Problem 7.

1. Find the ratio between the nuclear radius and the average electron radius in the $n = 1$ state, for H and for Pb. Use $R_{\text{nucleus}} \approx 1.2 A^{1/3}$ F, $A(\text{H}) = 1$, $A(\text{Pb}) = 208$.
2. Do the same for a muon ($M_\mu = 207 M_e$).
3. Is the picture of a pointlike nucleus reasonable in all these cases?

Problem 8. Replace $r^2 \to s$ in the radial equation of a harmonic oscillator potential. Find the changes in the constants $l(l+1)$, $M\omega^2$ and E that yield the Coulomb radial equation. Hint: make the replacement $R(r) \to s^{1/4}\Phi(s)$ and construct the radial equation using $s \equiv r^2$ as variable.

Problem 9. The positronium is a bound system of an electron and a positron (the same particle as an electron but with a positive charge). Their spin–spin interaction energy may be written as $\hat{H} = a\hat{\mathbf{S}}_\text{e}\cdot\hat{\mathbf{S}}_\text{p}$, where e and p denote the electron and positron, respectively. Obtain the energies of the resultant eigenstates (see Problem 11 of Chap. 5). Hint: apply a trick similar to the one used in Sect. 6.2.

Problem 10. Calculate the splitting between the $2p$ states with $m = 1/2$ of a hydrogen atom in the presence of spin–orbit coupling and a magnetic field B in the z-direction:

1. at the limit $v_{\text{so}} = 0$,
2. at the limit $B_z = 0$,
3. as a function of the ratio $q = 2\mu_\text{B} B_z/\hbar^2 v_{\text{so}}$.

Problem 11. Calculate the current associated with the spherical wave $A\exp(ikr)/r$ and show that the flux within a solid angle $d\Omega$ is constant.

Problem 12. Consider a planar motion.

1. What is the analogue of spherical symmetry in a two-dimensional space? Find the corresponding coordinates.
2. Write down the operator for the kinetic energy in these coordinates and find the degeneracy inherent in potentials with cylindrical symmetry.
3. Find the energies and degeneracies of the two-dimensional harmonic oscillator problem.
4. Verify that the function

$$\varphi_n = \frac{1}{x_c\sqrt{\pi n!}} \exp(-u^2/2) u^n \exp(\pm in\phi)$$

is an eigenstate of the Hamiltonian ($u = \rho/x_c$).

7 Many-Body Problems

7.1 The Pauli Principle

So far, we have discussed only one-particle problems. We now turn our attention to cases in which more than one particle are present. Here it is important to stress the fact that if $\hat{H} = \hat{H}(1) + \hat{H}(2)$, where $\hat{H}(1)$ and $\hat{H}(2)$ refer to different degrees of freedom (in particular, to different particles), and if $\hat{H}(1)\varphi_a(1) = E_a\varphi_a(1)$ and $\hat{H}(2)\varphi_b(2) = E_b\varphi_b(2)$, then

$$\hat{H}\varphi_a(1)\varphi_b(2) = (E_a + E_b)\varphi_a(1)\varphi_b(2) \ . \tag{7.1}$$

Let us now consider the case of two identical particles, 1 and 2. Two particles are identical if their interchange, in any physical operator, leaves the operator invariant:

$$[\hat{P}_{12}, \hat{Q}(1,2)] = 0 \ , \tag{7.2}$$

where $\hat{P}_{12}$ is the operator corresponding to the interchange process $1 \leftrightarrow 2$. As a consequence, the eigenstates of $\hat{Q}$ may be simultaneous eigenstates of $\hat{P}_{12}$ (Sect. 2.6). The operator $\hat{P}_{12}^2$ must have the single eigenvalue 1, since the system is left invariant by interchanging the particles twice. Thus the two eigenvalues of the operator $\hat{P}_{12}$ are ± 1. The eigenstates are said to be symmetric $(+1)$ or antisymmetric (-1) under the interchange of particles $1 \leftrightarrow 2$.

Consider two orthogonal single-particle states φ_p, φ_q which may in particular be eigenstates of a Hamiltonian. We construct the four two-body states by distributing the two particles in the two single-particle states. The symmetric combinations are

$$\Psi_{pp}^{(+)} = \varphi_p(1)\varphi_p(2) \ , \tag{7.3}$$

$$\Psi_{qq}^{(+)} = \varphi_q(1)\varphi_q(2) \ , \tag{7.4}$$

$$\Psi_{pq}^{(+)} = \frac{1}{\sqrt{2}}\left[\varphi_p(1)\varphi_q(2) + \varphi_q(1)\varphi_p(2)\right] \ , \tag{7.5}$$

while the antisymmetric state is

$$\Psi_{pq}^{(-)} = \frac{1}{\sqrt{2}}\left[\varphi_p(1)\varphi_q(2) - \varphi_q(1)\varphi_p(2)\right] \ . \tag{7.6}$$

The states (7.5) and (7.6) are called entangled states, meaning that they are not simply written as a component of the tensor product of the state vectors of particle 1 and particle 2 (see Sect. 10.1).

The average distance between two entangled identical particles is

$$\langle pq|(\mathbf{r}_1-\mathbf{r}_2)^2|pq\rangle_{(\pm)}^{1/2} = \left(\langle p|r^2|p\rangle + \langle q|r^2|q\rangle - 2\langle p|\mathbf{r}|p\rangle\langle q|\mathbf{r}|q\rangle \mp 2|\langle p|\mathbf{r}|q\rangle|^2\right)^{1/2}, \quad (7.7)$$

where the subscripts $(\pm)$ denote symmetric and antisymmetric states. The first three terms correspond to the average 'classical' distance which is obtained if state functions of the type $\varphi_p(1)\varphi_q(2)$ are used. According to (7.7), this classical distance may be decreased for entangled particles in symmetric states and increased if they are in antisymmetric states. Therefore, the symmetry induces correlations between identical particles, even in the absence of residual interacting forces.

We now generalize the construction of symmetric and antisymmetric states to ν identical particles. Let $\hat{P}_b$ denote the operator that performs one of the $\nu!$ possible permutations. It can be shown that this operator may be written as a product of two-body permutations $\hat{P}_{ij}$. Although this decomposition is not unique, the parity of the number η_b of such permutations is. We construct the operators

$$\hat{S} \equiv \frac{1}{\sqrt{\nu!}}\sum_b \hat{P}_b , \qquad \hat{A} \equiv \frac{1}{\sqrt{\nu!}}\sum_b (-1)^{\eta_b} \hat{P}_b . \quad (7.8)$$

Acting with the operator $\hat{S}$ on a state of ν identical particles produces a symmetric state, whilst acting with $\hat{A}$ produces an antisymmetric state.

A new quantum principle has to be added to those listed in Chap. 2:

Principle 4. *There are only two kinds of particles in nature:*[1] *bosons described by symmetric state vectors, and fermions described by antisymmetric state vectors.*

As long as the Hamiltonian is totally symmetric in the particle variables, its eigenstates may be labeled with their properties under the interchange of two particles (symmetry or antisymmetry). According to Principle 4, many otherwise possible states are eliminated. For instance, the only two-body fermion state that can be found in nature is (7.6).

---

[1] During the last twenty years it has been understood that, although this postulate holds true in our three-dimensional world, there is a whole range of intermediate possibilities – anyons – between bosons and fermions, in two dimensions. In some cases there are surface layers a few atoms thick in which the concept of anyons is realized [35], as in the fractional quantum Hall effect (Sect. 7.8.2*).

7.1 The Pauli Principle

All known particles with half-integer values of spin are fermions (electrons, muons, protons, neutrons, neutrinos, etc.). All known particles with integer spin are bosons[2] (photons, mesons, etc.).

Moreover, every composite object has a total angular momentum, which can be viewed as the composite particle spin, and which is obtained according to the addition rules of Sect. 5.3. If this spin has a half-integer value, the object behaves like a fermion, whereas a composite system with an integer value of the spin acts as a boson. For instance, He^3 is a fermion (two protons and one neutron), while He^4 is a boson (an α-particle, with two protons and two neutrons), in spite of the fact that both isotopes have the same chemical properties.

Let us distribute ν identical bosons into a set of single-particle states φ_p and denote by n_p the number of times that the single-particle state p is repeated. The n_p are called occupation numbers. In order to construct the symmetrized ν-body state vector, we start from the product

$$\Psi_{pq\ldots r}(1, 2, \ldots, \nu)$$
$$= \varphi_p(1)\varphi_p(2)\ldots\varphi_p(n_p)\varphi_q(n_p+1)\varphi_q(n_p+2)\ldots\varphi_q(n_p+n_q)\ldots\varphi_r(\nu)$$
$$= \varphi_p^{(n_p)}\varphi_q^{(n_q)}\ldots\varphi_r^{(n_r)}, \tag{7.9}$$

with $\sum_i n_i = \nu$. Subsequently the state vector is symmetrized by applying the operator $\hat{S}$. The final state is

$$\Psi_{n_p,n_q,\ldots,n_r}(1, 2, \ldots, \nu) = \mathcal{N}\hat{S}\Psi_{pq\ldots r}(1, 2, \ldots, \nu), \tag{7.10}$$

where $\mathcal{N}$ is a normalization constant. The occupation numbers label the states. There are no restrictions on the number of bosons in a given single-boson state. For instance, in the two-particle case, the possible symmetric state vectors are (7.3), (7.4) and (7.5).

We may also characterize the state by using the occupation numbers in the case of fermions. The procedure for constructing the antisymmetric state is the same, but for the application of the operator $\hat{A}$ instead of $\hat{S}$. However, the results are different in the sense that occupation numbers must be zero or one. Otherwise the state vector would not change its sign under the exchange of two particles occupying the same state. The antisymmetrization principle requires that fermions should obey Pauli's exclusion principle [37]: "If there is an electron in the atom for which these [four] quantum numbers have definite values, then the state is occupied, full, and no more electrons are allowed in."

The antisymmetric state function for ν fermions may be written as a Slater determinant

---

[2] Pauli produced a demonstration of this relation between spin and statistics which involved many complications of quantum field theory. Feynman's challenge that an elementary proof of the spin–statistics theorem be provided has not yet been answered [36].

$$\Psi_{pq...r}(1,2,\ldots,\nu) = \frac{1}{\sqrt{\nu!}} \begin{vmatrix} \varphi_p(1) & \varphi_p(2) & \cdots & \varphi_p(\nu) \\ \varphi_q(1) & \varphi_q(2) & \cdots & \varphi_q(\nu) \\ \vdots & \vdots & \ddots & \vdots \\ \varphi_r(1) & \varphi_r(2) & \cdots & \varphi_r(\nu) \end{vmatrix}. \qquad (7.11)$$

The permutation of two particles is performed by interchanging two columns, which produces a change of sign. All the single-particle states must be different. Otherwise, the two rows are equal and the determinant vanishes.

A widely used representation of the states (7.9) and (7.11), in terms of creation and annihilation operators, is given in Sect. 7.5*.

The possibility of placing many bosons in a single (symmetric) state gives rise to phase transitions, with important theoretical and conceptual implications that are illustrated for the case of the Bose–Einstein condensation (Sect. 7.7*). Even more spectacular consequences appear in the fermion case. Some of them will be treated later in this chapter.

7.2 Two-Electron Problems

Let us consider the case of the He atom. For the moment, we disregard the interaction between the two electrons. The lowest available single-particle states for the two electrons are the $\varphi_{100\frac{1}{2}m_s}$, $\varphi_{200\frac{1}{2}m_s}$ and $\varphi_{21m_l\frac{1}{2}m_s}$ states, where we use the same representation as in (6.7).

This problem involves four angular momenta: the two orbital and the two spin angular momenta. The two orbital angular momenta and the two spins may be coupled first[3] ($\hat{\boldsymbol{L}} = \hat{\boldsymbol{L}}_1 + \hat{\boldsymbol{L}}_2$ and $\hat{\boldsymbol{S}} = \hat{\boldsymbol{S}}_1 + \hat{\boldsymbol{S}}_2$). Subsequently, the addition of the total orbital and total spin angular momentum yields the total angular momentum $\hat{\boldsymbol{J}} = \hat{\boldsymbol{L}} + \hat{\boldsymbol{S}}$.

The spin part of the state vector may carry spin 1 or 0. We obtain the corresponding states $\chi^s_{m_s}$ by using the coupling given in (5.52), with $j_1 = j_2 = 1/2$. The (three) two-spin states with spin 1 are symmetric, while the state with spin 0 is antisymmetric. Thus,

$$\chi^1_1(1,2) = \varphi_\uparrow(1)\varphi_\uparrow(2)\,, \qquad \chi^1_0(1,2) = \frac{1}{\sqrt{2}}\left[\varphi_\uparrow(1)\varphi_\downarrow(2) + \varphi_\uparrow(2)\varphi_\downarrow(1)\right]\,,$$

$$\chi^1_{-1}(1,2) = \varphi_\downarrow(1)\varphi_\downarrow(2)\,, \qquad \chi^0_0(1,2) = \frac{1}{\sqrt{2}}\left[\varphi_\uparrow(1)\varphi_\downarrow(2) - \varphi_\uparrow(2)\varphi_\downarrow(1)\right]\,.$$

$$(7.12)$$

We now consider different occupation numbers for the two electrons.

---

[3] There is an alternative coupling scheme in which the orbital and spin angular momenta are first coupled in order to yield the angular momentum of each particle: $\hat{\boldsymbol{J}}_i = \hat{\boldsymbol{L}}_i + \hat{\boldsymbol{S}}_i$ ($i = 1, 2$), as in (6.8). Subsequently, the two angular momenta are coupled together: $\hat{\boldsymbol{J}} = \hat{\boldsymbol{J}}_1 + \hat{\boldsymbol{J}}_2$. The two coupling schemes give rise to two different sets of basis states.

1. The two electrons occupy the lowest orbit $\varphi_{100\frac{1}{2}m_s}$. In this case the spatial part is the same for both electrons, and thus the state vector is necessarily spatially symmetric. Therefore the symmetric spin state with spin 1 is excluded by the exclusion principle. Only the (entangled) state with zero spin can exist.
2. One electron occupies the lowest level $\varphi_{100\frac{1}{2}m_s}$ and the other the next level $\varphi_{200\frac{1}{2}m_s}$. In this case the difference in the radial wave functions allows us to construct both a symmetric and an antisymmetric state for the spatial part of the wave function [(7.5) and (7.6), respectively]. Both spatial states carry $l = 0$. Two total states are now allowed by the Pauli principle: the combination of the symmetric spatial part with the antisymmetric spin state and vice versa. The (so far neglected) interaction between the electrons breaks the degeneracy between these two allowed total states: according to (7.7), two electrons in a spatially antisymmetric state are further apart than in a symmetric state. They thus feel less Coulomb repulsion. Their energy decreases relative to the energy of the spatially symmetric state.
3. One electron occupies the lowest level $\varphi_{100\frac{1}{2}m_s}$ and the other the level $\varphi_{21m_l\frac{1}{2}m_s}$. It is left for the reader to treat this case as an exercise. He/she may follow the same procedure as in the previous example, bearing in mind that the orbital angular momentum no longer vanishes.

7.3 Periodic Tables

The attraction exerted by the nuclear center, proportional to Ze^2, allows us to implement a central-field description for atomic systems displaying more than one electron. However, the Hamiltonian also includes the Coulomb repulsion between electrons. This interaction is weaker (since it is only proportional to e^2), but an electron experiences $Z - 1$ such repulsions. We can nonetheless take them into account to a good approximation by modifying the central field because:

- electrons from occupied levels cannot be scattered to other occupied levels (Pauli principle), and when scattered to empty levels have to overcome the gap between the energy of the occupied level and the energy of the last filled state, thus reducing the effectiveness of the residual interaction;
- the electric fields created by electrons lying outside a radius r' tend to cancel for radius $r < r'$, due to the well-known compensation between the field intensity ($\propto 1/r^2$) and the solid angle ($\propto r^2$).

Although the optimum choice of the single-particle central potential constitutes a difficult problem, it is simple to obtain the behavior at the limits

$$\lim_{r \to 0} V(r) = -\frac{Ze^2}{4\pi\epsilon_0 r}, \qquad \lim_{r \to \infty} V(r) = -\frac{e^2}{4\pi\epsilon_0 r}. \qquad (7.13)$$

Fig. 7.1. Electronic shell structure. The figure gives a rough representation of the order of single-electron levels. *Numbers to the right* indicate the number of electrons in closed shell atoms

Close to the nucleus, the electron feels all the nuclear field. Far away, this field is screened by the remaining $Z-1$ electrons. The potential at intermediate points may be obtained qualitatively by interpolation.

The energy eigenvalues of this effective potential are also qualitatively reproduced by adding the term

$$\hat{H}_l = c\hat{L}^2 \tag{7.14}$$

to the Coulomb potential, since the centrifugal term $\hbar^2 l(l+1)/2Mr^2$ prevents the electrons occupying levels with large values of l from approaching the center, and thus feeling the greater attraction of the potentials (7.13).

The energies E_{nl} are also labeled by the orbital quantum number, since the potential is no longer simply proportional to $1/r$ (Sect. 6.1.1). They are qualitatively presented in Fig. 7.1, where the nomenclature of Table 5.2 is used. The set of energy levels which are close to each other is called a shell. In a closed shell, all magnetic substates are occupied.

The ground state of a given atom is determined by successively filling the different single-particle states until the Z electrons are exhausted. A closed shell carries zero orbital and zero spin angular momenta (see Problem 7). A closed shell displays neither loose electrons nor holes, and thus constitutes a quite stable system. This fact explains the properties of noble gases in the Mendeleev chart, for which $Z = 2, 10, 18, 36, 54$ and 86 (Fig. 7.1). The angular momenta (including the magnetic momenta), the degree of stability, the nature of chemical bonds, in fact, all the chemical properties, are determined by the outer electrons lying in the last, unfilled shell, a spectacular consequence of the Pauli principle.

The electron configuration of an atom with many electrons is specified by the occupation of the single-particle states of the unfilled shell. For instance, the lowest configuration in the Mg atom, with $Z = 12$, is[4] $(3s)^2$. Configurations $(3s)(3p)$ and $(3p)^2$ lie close in energy.

In the atomic case the total single-particle angular momentum j is not usually specified (as it was not in Sect. 7.2), because the strength of the spin-orbit coupling is small relative to the electron repulsions. However, for heavier elements and inner shells, the quantum numbers (l, j) become relevant once again.

Let us now consider the nuclear table. A nucleus has A nucleons, of which N are neutrons and Z are protons. In spite of the fact that there is no ab initio attraction from a nuclear center and the internucleon force is as complicated as it can be, the Pauli principle is still effective: the starting point for the description of most nuclear properties is a shell model. For systems with short range interactions, a realistic central potential follows the probability density, which in the nuclear case has a Woods–Saxon shape, $w(r)$. A strong spin–orbit interaction must be included on the surface, with an opposite sign, as in the atomic case. A central Coulomb potential also appears for protons:

$$\hat{V} = -v_0 w(r) - v_{\text{so}} \frac{r_0^2}{r} \frac{dw(r)}{dr} \hat{\boldsymbol{L}} \cdot \hat{\boldsymbol{S}} + V_{\text{coul}}$$

$$w(r) \equiv \left(1 + \exp\frac{r-R}{a}\right)^{-1}. \tag{7.15}$$

The empirical values of the parameters are [38]

$$v_0 = \left(-51 + 33\frac{N-Z}{A}\right) \text{ MeV} \quad \text{and} \quad v_{\text{so}} = 0.44 V_0 \, .$$

Here $a = 0.67$ F represents the skin thickness and $R = r_0 A^{1/3}$ is the nuclear radius, with $r_0 = 1.20$ F. The resulting shell structure is shown in Fig. 7.2.

Nucleons moving in a Woods–Saxon potential see a potential similar to the harmonic oscillator potential (Fig. 7.3). An attractive term of the form (7.14) should also be included, since the nuclear single-particle states in which nucleons lie close to the surface are more energetically favored by the Woods–Saxon potential than by the harmonic oscillator potential (Fig. 7.3). Therefore, the simpler effective potential

$$\hat{V} = \frac{M_p \omega^2}{2} r^2 - c \hat{\boldsymbol{L}} \cdot \hat{\boldsymbol{S}} - d \left(\hat{L}^2 - \langle L^2 \rangle_N\right) \tag{7.16}$$

may be used instead of (7.15), at least for bound nucleons (see Problem 3 of Chap. 6), where $\hbar\omega = 41$ MeV $A^{-1/3}$, $c = 0.13\hbar\omega$, and $d = 0.038\hbar\omega$ for

---

[4] The first number is the Coulomb principal quantum number; the orbital angular momentum follows the notation of Table 5.2; the exponent (2) denotes the number of particles with the previous two quantum numbers.

Fig. 7.2. Nuclear shell structure. This figure is an approximation of the order of single-nucleon levels. The numbers of nucleons for closed shell systems are indicated on the *right*

protons and $d = 0.024\hbar\omega$ for neutrons [38]. The symbol $\langle L^2 \rangle_N$ denotes the average value of L^2 in an N-oscillator shell (Problem 3 of Chap. 6). The eigenstates are labeled with the quantum numbers $Nljm\tau$, where the new quantum number τ equals $1/2$ for neutrons and $-1/2$ for protons.

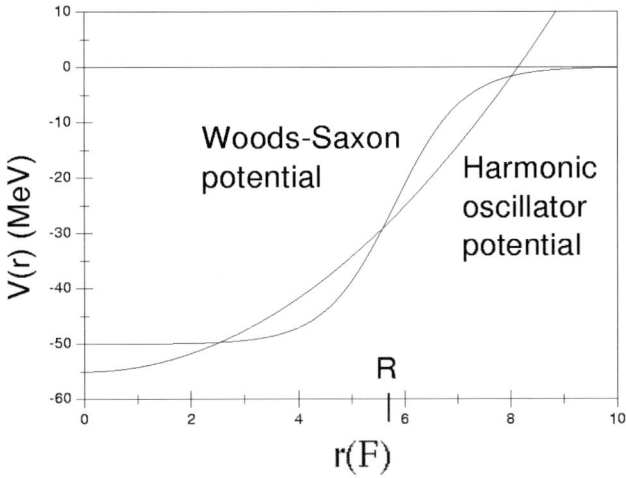

Fig. 7.3. Comparison of the Woods–Saxon and harmonic oscillator potentials [38]

The lowest shell $N = 0$ is filled up with four nucleons, two protons and two neutrons, giving rise to the very stable α-particle. As in the electron case, closed shells do not contribute to the properties of low-lying excited states. Note that both nucleons should fill closed shells in order to obtain the analogy of noble gases. This occurs in the nuclear systems $Z = N = 2$; $Z = N = 8$; $Z = N = 20$; $Z = 20, N = 28$; $Z = N = 28$; $Z = N = 50$; $Z = 50, N = 82$; and $Z = 82, N = 126$.

It should be stressed that, unlike the hydrogen case, the description of heavier atoms/nuclei in terms of a central field is, at best, a semi-quantitative approximation: one-body terms can never completely replace two-body interactions. The approximation is more reliable for systems that have one more particle (or hole) than a closed shell.

7.4 Motion of Electrons in Solids

7.4.1 Electron Gas

In the simplest possible model of a metal, electrons move independently of each other. The electrostatic attraction of the crystalline lattice prevents them from escaping when they approach the surface. The electron gas results of Sect. 4.4 may be easily generalized to the three-dimensional case. The wave states are given as the product of three one-dimensional solutions (4.41):

$$\varphi_{n_x n_y n_z} = \frac{1}{\sqrt{V}} \exp\left[i(k_{n_x} x + k_{n_y} y + k_{n_z} z)\right] . \tag{7.17}$$

The volume is $V = a^3$. The allowed $\boldsymbol{k}$ values constitute a cubic lattice in which two consecutive points are separated by the distance $2\pi/a$ (4.40)

$$k_{n_i} = \frac{2\pi}{a} n_i , \quad n_i = 0, \pm 1, \pm 2, \ldots , \quad i = x, y, z . \tag{7.18}$$

The energy of each level is

$$\epsilon_k = \frac{\hbar^2 |\boldsymbol{k}|^2}{2M} . \tag{7.19}$$

In order to build the ν-electron ground state, we start by putting two electrons on the level $k_x = k_y = k_z = 0$. We successively fill the unoccupied levels as their energy increases. When there is a large number of electrons, the occupied region will be indistinguishable from a sphere in k-space. The radius of this sphere is called k_F, the Fermi momentum, and its energy $\epsilon_F \equiv \hbar^2 k_F^2/2M$, the Fermi energy. At zero energy, the levels with $|\boldsymbol{k}| \leq k_F$ are occupied pairwise and those above are empty. Since we are interested in the large-volume limit, the levels are very close together and we may replace summations with integrals that have a volume element (4.43). Thus,

$$\sum_k f_k \approx \frac{V}{8\pi^3} \int f_k \mathrm{d}^3 k \ . \tag{7.20}$$

An electron gas is characterized by the Fermi temperature $T_\mathrm{F} \equiv \epsilon_\mathrm{F}/k_\mathrm{B}$, where k_B is the Boltzmann constant (Table 13.1). If the temperature $T \ll T_\mathrm{F}$, the electron gas has properties that are very similar to the $T = 0$ gas. The number of levels per unit volume with energy less than ϵ and the density of states per unit interval of energy per unit volume are

$$n(\epsilon) = \frac{2}{V} \sum_{n_x n_y n_z} \approx \frac{1}{4\pi^3} \int_{k \leq k_\epsilon} \mathrm{d}^3 k = \frac{k_\epsilon^3}{3\pi^2} = \frac{1}{3\pi^2}\left(\frac{2M\epsilon}{\hbar^2}\right)^{\frac{3}{2}} ,$$

$$\rho(\epsilon) = \frac{\partial n}{\partial \epsilon} = \frac{1}{2\pi^2}\left(\frac{2M}{\hbar^2}\right)^{\frac{3}{2}} \epsilon^{\frac{1}{2}} , \tag{7.21}$$

respectively. At the Fermi energy, the value of these quantities is

$$n_\mathrm{F} = \frac{1}{3\pi^2} k_\mathrm{F}^3 , \qquad \rho_\mathrm{F} = \frac{3 n_\mathrm{F}}{2 \epsilon_\mathrm{F}} . \tag{7.22}$$

For the Na typical case: $n_\mathrm{F} \approx 2.65 \times 10^{22}$ electrons/cm^3, $k_\mathrm{F} \approx 0.92 \times 10^8$ cm^{-1}, $\epsilon_\mathrm{F} \approx 3.23$ eV, and $T_\mathrm{F} \approx 3.75 \times 10^4$ K.

We now explore some thermal properties of an electron gas. If the electrons obeyed classical mechanics, each of them would gain an energy of the order of $k_\mathrm{B} T$ in going from absolute zero to the temperature T. The total thermal energy per unit volume of electron gas would be of the order of

$$u_\mathrm{cl} = n_\mathrm{F} k_\mathrm{B} T , \tag{7.23}$$

and the specific heat at a constant temperature would thus be independent of the temperature:

$$(C_V)_\mathrm{class} = \frac{\partial u_\mathrm{cl}}{\partial T} = n_\mathrm{F} k_\mathrm{B} . \tag{7.24}$$

However, the Pauli principle prevents most of the electrons from gaining energy. Only those with an initial energy ϵ_k such that $\epsilon_\mathrm{F} - \epsilon_k < k_\mathrm{B} T$ can be expected to gain energy. The number of such electrons is given roughly by

$$\rho_\mathrm{F} k_\mathrm{B} T = \frac{3 n_\mathrm{F}}{2} \frac{T}{T_\mathrm{F}} . \tag{7.25}$$

The total thermal energy and specific heat per unit volume are

$$u = \rho_\mathrm{F} (k_\mathrm{B} T)^2 , \tag{7.26}$$

$$C_V = 3 n_\mathrm{F} k_\mathrm{B} \frac{T}{T_\mathrm{F}} . \tag{7.27}$$

The specific heat is proportional to the temperature and is reduced by a factor $\approx 1/30$ at room temperature.

The probability of an electron being in a state i of energy ϵ_i is given by the Fermi–Dirac distribution $n(\epsilon)$ (7.39). Using this distribution, the expression for the total energy per unit volume is

$$u = \frac{1}{4\pi^3} \int \epsilon_k n(\epsilon_k) \mathrm{d}^3 k \,, \tag{7.28}$$

which is a better approximation than (7.26). Upon integration, one obtains results similar to (7.27).

7.4.2* Band Structure of Crystals

Although the electron gas model explains many properties of solids, it fails to account for electrical conductivity, which can vary by a factor of 10^{30} between good insulators and good conductors.

A qualitative understanding of conductors and insulators may be obtained from a simple generalization of the band model described in Sect. 4.7*. As a consequence of the motion of single electrons in a periodic array of ions, the possible individual energies are grouped into allowed bands. Each band contains $2N$ levels, where N is the number of ions, and the factor 2 is due to spin.

According to the Pauli principle, we obtain the ground state by successively filling the individual single-particle states of the allowed bands. The filled bands are called valence bands. If we place the solid within an electric field, the electrons belonging to a valence band cannot be accelerated by a small electric field, since they would tend to occupy other states of the same band, which are already occupied. Much like the case of closed shells in atoms and nuclei, the electrons in the valence band constitute inert systems which do not contribute to thermal or electrical properties. A solid consisting only of valence bands is an insulator. The insulation gets better as the distance ΔE between the upper valence band and the next (empty) band increases.

By contrast, electrons in partially filled bands can easily absorb energy from an applied electric field. Such a band is called a conduction band.

The previous considerations are valid for $T = 0$. In solids which are insulators at $T = 0$, the thermal motion increases the energy of the electrons by an amount $k_\mathrm{B} T$. If ΔE is of the order of $k_\mathrm{B} T$, some electrons belonging to the last valence band may jump to the conduction band. This system is a semiconductor. The conductivity varies as $\exp(-\Delta E/k_\mathrm{B} T)$.

Now we see that the existence of conductors, semiconductors and insulators is a consequence of the Pauli principle. Another consequence arises from the fact that the electrons which jump to the conduction band leave empty states called holes in the valence band. Other electrons of the same valence band may occupy these holes, leaving other holes behind them. Thus there is a current, due to the electrons of the valence band, which is produced by the holes. The holes carry a positive charge, because they represent the absence of an electron.

7.5* Occupation Number Representation

The representation (3.43) of the harmonic oscillator states may be straightforwardly generalized to the case in which ν oscillators are present. The eigenstates and eigenvalues of the energy are [see (7.1)]

$$\varphi_{n_1,n_2,\ldots,n_\nu} = \prod_{p=1}^{p=\nu} \frac{1}{\sqrt{n_p!}} (a_p^+)^{n_p} \varphi_0 , \qquad a_p \varphi_0 = 0 ,$$

$$E_{n_1,n_2,\ldots,n_\nu} = \sum_{p=1}^{p=\nu} E_p n_p , \qquad (7.29)$$

where n_p is an eigenvalue of the operator $\hat{n}_p = a_p^+ a_p$ ($n_p = 0, 1, 2, \ldots$). We have disregarded the ground state energy.

The creation and annihilation operators corresponding to different subscripts commute with each other:

$$[a_p, a_q] = [a_p^+, a_q^+] = 0 , \qquad [a_p, a_q^+] = \delta_{pq} . \qquad (7.30)$$

The n_p quanta that occupy the p-state are indistinguishable from each other. They are therefore bosons, and states $\varphi_{n_1,n_2,\ldots,n_\nu}$ constitute another representation of states (7.10). An example of the occupation number representation is given by the quantized radiation field (Sect. 9.5.2).

The matrix elements $\langle q|Q|p\rangle$ of an operator $\hat{Q}$ between single-particle states are reproduced within the occupation number representation, where the operator reads $\hat{Q} = \sum_{q,p} \langle q|Q|p\rangle a_q^+ a_p$:

$$\langle n_1, n_2 \ldots (n_q+1), (n_p-1) \ldots n_\nu |Q| n_1, n_2 \ldots n_q, n_p \ldots n_\nu\rangle \qquad (7.31)$$
$$= \langle (n_q+1), (n_p-1)|Q|n_q, n_p\rangle$$
$$= \langle q|Q|p\rangle \sqrt{(n_q+1)(n_p)}$$
$$= \langle q|Q|p\rangle \quad \text{if} \quad n_p = 1, \; n_q = 0 .$$

A similar equivalence may be obtained for n-body operators.

The operators $a_{lm_l}^+$ have the same coupling properties (5.32) as the spherical harmonics Y_{lm_l}.

It is natural to seek a similar formalism which may apply to fermions. A system of such particles can be described as a many-particle state vector that changes its sign with the interchange of any two particles. Since the required linear combination of products of one-particle states [Slater determinant (7.11)] can be uniquely specified by listing the singly occupied states, the formalism we seek must limit the eigenvalues of $\hat{n}_p$ to 0 and 1.

The desired modification consists in the replacement of commutators $[\hat{A}, \hat{B}]$ by anticommutators

$$\{\hat{A}, \hat{B}\} \equiv \hat{A}\hat{B} + \hat{B}\hat{A} . \qquad (7.32)$$

The anticommutators of creation and annihilation operators read

$$\{a_p, a_q\} = \{a_p^+, a_q^+\} = 0 , \qquad \{a_p^+, a_q\} = \delta_{pq} . \tag{7.33}$$

The eigenvalues of the number operators are obtained by constructing the operator equation

$$(\hat{n}_p)^2 = a_p^+ a_p a_p^+ a_p = a_p^+(1 - a_p^+ a_p)a_p = a_p^+ a_p = \hat{n}_p . \tag{7.34}$$

Since both operators $(\hat{n}_p)^2$ and $\hat{n}_p$ are simultaneously diagonal, (7.34) is equivalent to the algebraic equation $n_p^2 = n_p$, which has two roots: 0 and 1. Fermions thus obey the exclusion principle. Within this limitation, the eigenstates and energies (7.29) are also valid for the case of fermions. Note that an interchange of two fermion creation operators changes the sign of the eigenststate [as happens for the Slater determinant (7.11)].

Operators within the fermion number representation are also constructed as in (7.31). When acting on products of fermion states (7.29), care must be taken with the number of permutations between creation and annihilation operators in order to obtain consistent phases.

The operators a_{lsjm}^+ have coupling properties (5.32) identical to the states φ_{lsjm}.

7.6* Quantum Statistics

Differences between counting the number of states according to whether the particles are distinguishable or not and, in the latter case, whether they are bosons or fermions, have already appeared in the two-body case, as shown in Sect. 7.1. If three particles have to be distributed into three states, we may construct:

- one antisymmetric Slater determinant (7.11),
- 10 symmetric states [three states $\varphi_a^{(3)}$, six states $\varphi_a^{(2)}\varphi_b^{(1)}$ and one state $\varphi_a^{(1)}\varphi_b^{(1)}\varphi_c^{(1)}$ (7.9)],
- sixteen states which are neither symmetric nor antisymmetric.

Such differences lead to different occupation probabilities $n(\epsilon)$ for the levels with energy ϵ. Let us assume that:

1. The equilibrium distribution is the most probable distribution consistent with a constant number of particles and a constant energy.
2. The particles are identical.
3. The particles are distinguishable.
4. There is no restriction on the number of particles in any state.

Given these assumptions, one derives the classical Maxwell–Boltzmann distribution [Fig. 7.4(M-B)]:

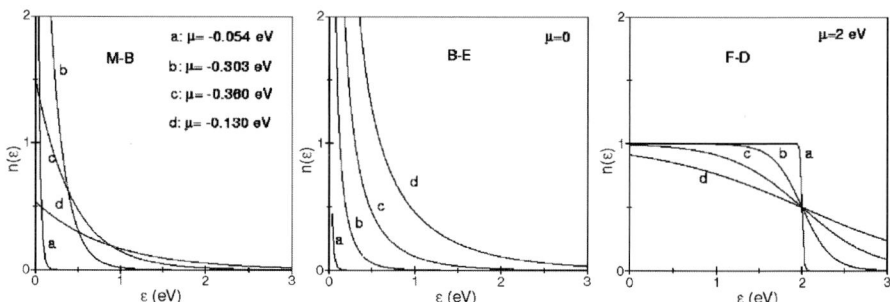

Fig. 7.4. Maxwell–Boltzmann (M-B), Bose–Einstein (B-E) and Fermi–Dirac (F-D) distribution functions. The value $n(\epsilon)$ gives the fraction of levels at a given energy which are occupied when the system is in thermal equilibrium. The curves correspond to: (**a**) $T = 300$ K, (**b**) $T = 1000$ K, (**c**) $T = 5000$ K and (**d**) $T = 10\,000$ K

$$n(\epsilon) = \exp\left(-\frac{\epsilon - \mu}{k_B T}\right). \tag{7.35}$$

where μ is a constant fixing the number of particles. For a system of harmonic oscillators, $\mu = -k_B T \log k_B T$.

In the quantum mechanical case, this distribution can only hold if the particles do not overlap. If this is not the case and, consequently, assumption 3 is removed, one obtains the Bose–Einstein distribution [Fig. 7.4(B-E)], which applies to bosons [39]

$$n(\epsilon) = \left[\exp\left(\frac{\epsilon - \mu}{k_B T}\right) - 1\right]^{-1}. \tag{7.36}$$

If we choose zero for the energy of the ground state, the occupancy of the ground state is

$$n(0) = \left[\exp\left(-\frac{\mu}{k_B T}\right) - 1\right]^{-1}. \tag{7.37}$$

In the limit $T \to 0$, the occupancy of the ground state equals the total number of particles N:

$$\lim_{T \to 0} n(0) = \lim_{T \to 0} \left(1 - \frac{\mu}{k_B T} + \cdots - 1\right)^{-1} \approx -\frac{k_B T}{\mu} \approx N, \qquad \mu = -\frac{k_B T}{N}, \tag{7.38}$$

as $T \to 0$. Thus, in a boson system, the constant μ must lie below the ground state energy, if the occupations are to be non-negative numbers. In Fig. 7.4(B-E) it is assumed that $k_B T \ll N$

Moreover, if assumption 4 above is replaced by the condition that the number of particles in each level may be 0 or 1, the Fermi–Dirac distribution is derived [Fig. 7.4(F-D)] [40, 41]:

$$n(\epsilon) = \left[\exp\left(\frac{\epsilon - \mu}{k_\mathrm{B} T}\right) + 1\right]^{-1}. \tag{7.39}$$

In the case of an electron gas, the parameter μ may be approximated by the Fermi energy ϵ_F (Sect. 7.4.1). Thus $\mu \approx 3.23$ eV for the Na case.

For values of $(\epsilon - \mu)/k_\mathrm{B} T \gg 1$, the three distributions coincide.

7.7* Bose–Einstein Condensation

Below a certain critical temperature T_c, a substantial portion of the total number of particles in a system of non-interacting bosons will occupy the single orbital of the lowest energy, essentially coming to rest. Any other single orbital, including the orbital of the second lowest energy, will be occupied by a relatively negligible number of particles. This effect is called Bose–Einstein condensation. Its experimental realization was made possible by the development of techniques for cooling, trapping and manipulating atoms.[5] (The sources [42] have been used for this section.)

In Einstein's original prediction, all bosons were supposed to be slowed down to zero momentum, which implies a macroscopic space, according to the uncertainty principle. However, any experimental setup requires some confinement. An important feature characterizing the available magnetic traps for alkali atoms is that the confining potential can be safely approximated by the quadratic form

$$V_\mathrm{ext}(\boldsymbol{r}) = \frac{M\omega^2}{2} r^2. \tag{7.40}$$

Neglecting the interaction between atoms implies that the Hamiltonian eigenvalues have the form (6.28). Thus we are considering a system composed of a large number ν of non-interacting bosons. The many-body ground state $\varphi(\boldsymbol{r}_1, \ldots, \boldsymbol{r}_\nu)$ is obtained by putting all the particles into the lowest single-particle state $\varphi_0(\boldsymbol{r})$:

$$\varphi(\boldsymbol{r}_1, \ldots, \boldsymbol{r}_\nu) = \prod_{i=1}^{i=\nu} \varphi_0(\boldsymbol{r}_i), \quad \varphi_0(\boldsymbol{r}) = \left(\frac{M\omega}{\pi\hbar}\right)^{3/4} \exp(-r^2/2x_\mathrm{c}^2). \tag{7.41}$$

The density distribution then becomes $\rho(\boldsymbol{r}) = \nu|\varphi_0(\boldsymbol{r})|^2$. While its value grows with ν, the size of the cloud is independent of ν and is fixed by the harmonic oscillator length x_c (3.33). It is typically of the order of $x_\mathrm{c} \approx 1$ µm in today's experiments.

---

[5] Because of the requirement of very low temperatures, nobody attempted to find this condensation until Fritz London did so in 1938. However, his interpretation of the He^4 transition at 2.4 K as a transition to the Bose–Einstein condensate is marred by the presence of large residual interactions [43].

At finite temperatures, particles are thermally distributed among the available states. A rough estimate may be obtained by assuming $k_B T \gg \hbar\omega$. In this limit we may use a classical Boltzmann distribution

$$n(\boldsymbol{r}) = \exp\left(-M\omega^2 \boldsymbol{r}^2/2k_B T\right) , \qquad (7.42)$$

which displays the much broader width

$$\frac{k_B T}{M\omega^2} = x_c \frac{k_B T}{\hbar\omega} \gg x_c . \qquad (7.43)$$

Therefore, the Bose–Einstein condensation in harmonic traps appears as a sharp peak in the central region of the density distribution. Figure 7.5 displays the density for 5000 non-interacting bosons in a spherical trap at temperature $T = 0.9 T_c$, where T_c is the critical temperature (see below). The central peak is the condensate, superimposed on the broader thermal distribution. Distance and density are in units of x_c and x_c^{-2}, respectively. The density is normalized to the number of atoms [42].

The momentum distribution of the condensate is also Gaussian, having a width $\hbar/x_c$ (3.51). The momentum distribution of the thermal particles is broader, of the order of $(k_B T)^{1/2}$. In fact these two momentum distributions are also represented by the curves in Fig. 7.5, provided the correct units are substituted.

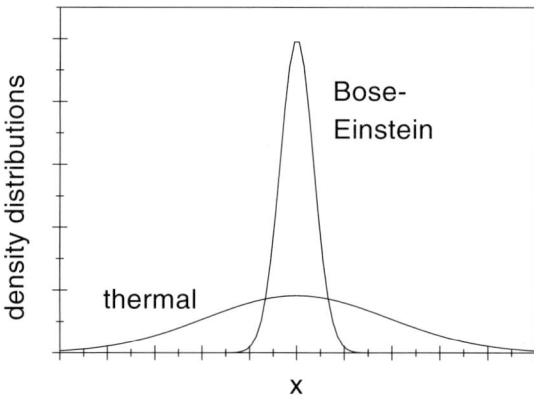

Fig. 7.5. Comparison between the densities of the condensate and the thermal particles

In 1995, rubidium atoms confined within a magnetic trap were cooled to the submicrokelvin regime by laser methods and then by evaporation [45]. The trap was suddenly turned off, allowing the atoms to fly away. By taking pictures of the cloud after various time delays, a two-dimensional momentum distribution of the atoms was constructed. As the temperature was lowered, the familiar Gaussian hump of the Maxwell–Boltzmann distribution was pierced by a rapidly rising sharp peak caused by atoms in the ground state of the trap, that is, by the condensate.

By allowing the trap to have cylindrical symmetry, the average momentum along the short axis was double that along the other [see the harmonic oscillator predictions (3.51)]. In contrast, the momentum distribution is always isotropic for a classical gas, unless it is flowing.

The calculation of the critical temperature T_c involves concepts of statistical mechanics that lie beyond the scope of this discussion. The experimental results for the condensate fraction closely follow the thermodynamic limit [45]:

$$\frac{T}{T_c} = 1 - \left(\frac{\nu_0}{\nu}\right)^{1/3}.$$

For 40 000 particles, T_c is approximately 3×10^{-7} K.

The first demonstration of the Bose–Einstein condensation involved 2000 atoms. Today millions are being condensed.

Bose–Einstein condensation is unique because it is the only pure quantum mechanical phase transition: it takes place without any interaction between the particles. This field is presently full of activity: collective motion, condensation and damping times of the condensate, its interaction with light, collision properties, and effects of the residual interactions are just some of the themes that are currently under very intense theoretical and experimental study.

7.8* Quantum Hall Effects

A planar sample of conductive material is placed in a magnetic field perpendicular to its surface. An electric current I is made to pass from one end to the other by means of a potential V_L. The longitudinal resistivity is the ratio $R_L = V_L/I$. Because of the Lorentz force, more electrons accumulate on one side of the sample than on the other, thereby producing a measurable voltage V_H – the Hall voltage – across the sample. The ratio $R_H = V_H/I$ is called the Hall resistivity. It increases linearly with the magnetic field.

However, in 1980 Klaus von Klitzing found that, for samples cooled to within one degree kelvin and placed in strong magnetic fields, the Hall resistivity exhibits a series of plateaus, i.e., intervals in which the Hall resistivity appears not to vary at all with the magnetic field [46]. Figure 7.6 displays a diagram of the measured inverse Hall resistance $h/e^2 R_H$ as a function of the density of electrons n times the characteristic area of the problem (7.47). The longitudinal resistance is sketched as well. Where the Hall resistivity is constant, the longitudinal resistivity practically vanishes. Moreover, the resistivity is $R_H = h/e^2 n$, with an amazing accuracy of order 10^{-6}. Here h is Planck's constant, e is the electron charge and n is an integer (Fig. 7.6). This is even true for samples with different geometries and with different processing histories, as well as for a variety of materials. It is the integer quantum Hall effect.

In 1982 Daniel Tsui, Horst Störmer and Arthur Gossard discovered other plateaus at which n has specific fractional values (1/3, 2/5 and 3/7) [47]. This is the fractional quantum Hall effect.

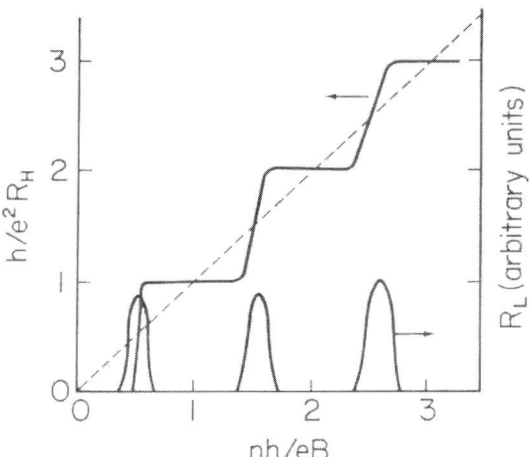

Fig. 7.6. The integer quantum Hall effect appears as plateaus in the Hall resistivity of a sample which coincide with the disappearance of the sample's electrical resistivity, as the magnetic field strength is varied [48]. (Reproduced with permission from Springer-Verlag)

7.8.1* Integer Quantum Hall Effect

Consider the planar motion of an electron in an external, uniform, magnetic field perpendicular to the plane.[6] For the sake of simplicity we will study circular geometries, and thus choose the symmetric gauge $A_x = yB/2$, $A_y = -xB/2$. Replacing the momentum $\hat{\boldsymbol{p}}$ by the effective momentum $\hat{\boldsymbol{p}} - e\boldsymbol{A}$ [50] in the free particle Hamiltonian yields

$$\hat{H} = \frac{1}{2M}\hat{\boldsymbol{p}}^2 + \frac{e^2 B^2}{8M}\rho^2 + \frac{\mu_B B}{\hbar}\left(\hat{L}_z + 2\hat{S}_z\right) . \tag{7.44}$$

The first two terms in (7.44) represent a two-dimensional harmonic Hamiltonian. The eigenvalues and eigenstates of the two-dimensional harmonic oscillator are (see Problem 12 in Chap. 6):

---

[6] The main source of this section is reference [49].

7.8* Quantum Hall Effects

$$E_n = \hbar\omega(n+1), \quad n = 0, 1, 2, \ldots, \quad x_c = \sqrt{\hbar/M\omega},$$

$$\varphi_{nm_l}(\rho,\phi) = R_{n_\rho m_l}(u)\varphi_{m_l}(\phi), \quad \rho = \sqrt{x^2+y^2}, \quad \phi = \tan^{-1}\frac{y}{x}, \quad u = \rho/x_c,$$

$$\varphi_{m_l}(\phi) = \frac{1}{\sqrt{2\pi}}\exp(im_l\phi), \quad m_l = n, n-2, \ldots, -n,$$

$$R_{n_\rho m_l}(u) = N_{nm_l}\exp(-u^2/2)u^{m_l}f_{n_\rho m_l}(u^2), \quad n_\rho = \frac{1}{2}(n - m_l).$$

The first two terms in (7.44) yield the frequency

$$\omega = -\frac{eB}{2M} = \frac{\mu_B B}{\hbar}. \tag{7.45}$$

The term proportional to $\hat{L}_z$ arises from the cross product of the square of the effective momentum, and the term proportional to $\hat{S}_z$ arises from (5.25). Since the change in energy produced by increasing $\hbar\omega$ by one unit is exactly compensated by decreasing the orbital angular momentum by one unit of $\hbar$, there are sets of degenerate states called Landau levels. In particular, the lowest Landau level is made up of radial nodeless states and values of $m_l = -n$. It has zero energy, since the spin term ($s_z = -\hbar/2$) compensates exactly for the zero point energy $\hbar\omega$ of the harmonic oscillator.

To keep the electron from flying apart, one adds a radial confining potential which does not alter the symmetry of the problem. Let us assume that all states of the first Landau level are occupied up to and including a maximum angular momentum $|M_l|$. The expectation value of the density,

$$\sum_{n=0}^{n=M_l}|\varphi_{n(-n)}|^2 = \frac{2}{x_c^2}\exp(-u^2)\sum_{n=0}^{n=M_l}\frac{1}{n!}u^{2n}, \tag{7.46}$$

is constant for $u \equiv \rho/x_c \ll \sqrt{M_l}$ and drops rapidly to zero around $\sqrt{M_l}$. This configuration is incompressible, since a compression would require the promotion of an electron to a higher Landau level.

Since the characteristic area of the problem is

$$\frac{\hbar\pi}{M\omega} = \frac{h}{|e|B}, \tag{7.47}$$

the constant of proportionality between the number of electrons per unit area and the strength of the magnetic field depends only on Planck's constant and the charge of the electron.

However, there are impurities and, consequently, the single energy of a Landau level is spread out into a band. The states of the band belong to two classes: states near the bottom or the top of the band are localized states.[7]

---

[7] Low (high) energy localized states arise around impurity atoms which have an excess (dearth) of positive charge.

Near the center of each energy band are extended states, each one spreading out over a large space. They are the only ones that may carry currents.

At very low temperatures, only states below the Fermi energy are occupied. Assume that the Fermi level is in the sub-band of localized states near the top of some Landau band. Now gradually increase the strength of the magnetic field, adjusting the current so that the Hall voltage remains constant. Since the number of states per unit area is proportional to the applied magnetic field [see (7.47)], the number of levels in each Landau state increases proportionately.

Many of the newly available states will be below the Fermi level, so electrons from higher-energy localized states will drop to fill them. As a result, the Fermi level descends to a lower position. However, as long as it remains in the sub-band of high-energy localized states, all the extended states remain fully occupied. The amount of current flowing therefore remains constant. Although the increased magnetic field slows the forward motion of any current-carrying electrons, this effect is compensated by the increase in the number of electrons in extended states.

As the Fermi level descends through the sub-band of extended states, some of them are vacated. The amount of current flowing decreases, while the Hall resistivity increases.

Eventually, the extended states will be emptied and the Fermi level will enter the sub-band of low-energy localized states. If there is at least one full Landau band below the Fermi level, the extended states in that band will be able to carry a constant current. However, because the extended states in one band have been completely emptied, the number of sub-bands of extended states has been reduced by one, and the Hall resistance is larger than it was on the previous plateau. The current is proportional to the number of occupied sub-bands of extended states, and on each plateau an integral number of these sub-bands is filled.

The second striking feature of the quantized Hall effect is that current flows without resistance in the plateau region. Recall that, to dissipate power, an electron must make a transition to a state of lower energy, the excess energy being distributed within the lattice as vibrations or heat.

First we should examine the regime between two plateaus. The Fermi energy varies slightly from point to point: the voltage measured at opposite sides of the sample gauges the difference between the Fermi energies at the two points. Thus, an electron can find itself in an extended state that is below the local Fermi energy in one region of space, but which extends into a region where its energy is above the local Fermi energy. The electron would thus be able to drop into a lower level, dissipating some energy into the lattice. This sample exhibits electrical resistance.

If the Fermi energy is within the region of energy corresponding to localized states, there may also be empty states of lower energy. These states, however, are far apart in space, at distances much larger than the localization distance. Electrons cannot drop to lower states, and thus they cannot dissipate energy.

Within this model, localized states also act as a reservoir of electrons, so that, for finite ranges of magnetic field strengths, the extended states in each Landau band are either completely empty or completely filled. Without the reservoir, the width of the regions displaying an integer quantum Hall effect would be vanishingly small.

Therefore, a relatively simple model of independent electrons, moving under the influence of electric and magnetic fields, can account for the main properties of the integer quantum Hall effect. This model has some features in common with the band structure of metals that explains the existence of conductors and insulators (Sect. 7.4.2*).

The quantized Hall effect enables us to calibrate instruments with extreme accuracy as well as to measure fundamental physical constants more precisely than ever before.

7.8.2* Fractional Quantum Hall Effect

The fractional quantum Hall effect is seen only when a Landau level is partially filled. For instance, a plateau is seen when the lowest Landau level is approximately 1/3 full. In this case the Hall resistivity is equal to 1/3 of the square of the electron charge divided by Planck's constant.

The independent particle model previously used to explain the integer quantum Hall effect does not show any special stability when a fraction of the states is filled. To explain the fractional quantum Hall effect, we must consider the interaction between electrons (like everything associated with open shells).

It is helpful to write the total wave function of particles moving in a Landau level as a Slater determinant (7.11):

$$\Psi = \left(\frac{\sqrt{2}}{x_c}\right)^{M_l+1} \frac{1}{\sqrt{(M_l+1)!}} \left(\prod_{n=0}^{n=M_l} \frac{1}{\sqrt{n!}}\right) \Phi(1, 2, \ldots, M_l+1), \quad (7.48)$$

where

$$\Phi = \exp\left(-\frac{1}{2}\sum_{i=1}^{i=M_l+1} |z_i|^2\right) \begin{vmatrix} 1 & 1 & \cdots & 1 \\ z_1 & z_2 & \cdots & z_{M_l+1} \\ z_1^2 & z_2^2 & \cdots & z_{M_l+1}^2 \\ \vdots & \vdots & \ddots & \vdots \\ z_1^{M_l} & z_2^{M_l} & \cdots & z_{M_l+1}^{M_l} \end{vmatrix}$$

$$= \exp\left(-\frac{1}{2}\sum_{i=1}^{i=M_l+1} |z_i|^2\right) \prod_{i>j=0}^{i>j=M_l-1} (z_i - z_j). \quad (7.49)$$

Here $z \equiv u \exp(-\mathrm{i}\phi)$.

In 1983, Robert Laughlin modified this wave function as follows [51]:

$$\Phi_\nu = \exp\left(-\frac{1}{2}\sum_{i=1}^{i=M_l+1}|z_i|^2\right)\prod_{i>j=0}^{i>j=M_l-1}(z_i-z_j)^\nu. \tag{7.50}$$

These wave states are exact ground states in the limit when electron–electron repulsions become infinitely short-ranged. The exponent ν measures the fraction of filled states: the wave functions have the required stability when ν equals 1/3, 1/5, 1/7, 2/3, 4/5 or 6/7. The denominator in the fraction ν must be an odd number in order to satisfy the Pauli principle. A mechanism based on localized and extended states may also be invoked here but, instead of being applied to independent electrons, it must be used for quasi-particles, which may be described as fractionally charged anyons (see the footnote on p. 96). The topic of anyons lies beyond the scope of this text.

Problems

Problem 1. Two particles with equal mass M are confined by a one-dimensional harmonic oscillator potential characterized by the length x_c. Assume that one is in the eigenstate $n = 0$ and the other in $n = 1$. Find the probability density for the relative distance $x = x_a - x_b$, the root mean square value of x, and the probability of finding the two particles within a distance of $x_c/5$ from each other if they are:

1. non-identical particles,
2. identical bosons,
3. identical fermions.

Hint: write the two-particle wave function in terms of the relative coordinate x and the center of mass coordinate $x_g = (x_a + x_b)/2$ and integrate the total probability density with respect to x_g.

Problem 2. Consider a He atom in which one electron is in the state $\varphi_{100\frac{1}{2}m_s}$ and the other in the state $\varphi_{21m_l\frac{1}{2}m_s}$.

1. Construct the possible two-electron states.
2. Split the energy of the allowed states in a qualitative manner by including a Coulomb repulsion between the electrons.

Problem 3. Couple two independent bosons, each carrying spin two. What spin angular momenta are possible? [See the relations (5.34)].

Problem 4. State whether the spatial sector of a two-body vector state is symmetric or antisymmetric with respect to the interchange of the particles, if the spin sector is given by:

1. $[\varphi_{\frac{1}{2}}(1)\,\varphi_{\frac{1}{2}}(2)]^0$,
2. $[\varphi_{\frac{1}{2}}(1)\,\varphi_{\frac{1}{2}}(2)]^1$,
3. $[\varphi_1(1)\,\varphi_1(2)]^0$,
4. $[\varphi_1(1)\,\varphi_1(2)]^1$,
5. $[\varphi_1(1)\,\varphi_1(2)]^2$.

Problem 5. What angular momenta are possible for two fermions constrained to move in a j-shell? A j-shell is constituted by the set of states which have the same quantum numbers, including j, with the exception of the projection m.

Problem 6. Couple the spin states of a deuteron ($s_\mathrm{d} = 1$) and a proton ($s_\mathrm{p} = 1/2$). What total spins are possible:

1. if we ignore the Pauli principle?
2. if the three nucleons move within the $N = 0$ harmonic oscillator shell and the Pauli principle is taken into account? (This is approximately the He^3 ground state.)

Problem 7. Show that a closed fermion j-shell carries zero angular momentum:

1. using the Slater determinant (7.11),
2. using the occupation number representation (7.29) plus the anticommutation relations (7.33).

Problem 8. Calculate the total number of protons $\mathcal{N}$, the angular momentum j and the parity of nuclei with one proton more than a closed nuclear j-shell. Assume that the Hamiltonian used in Problem 3 of Chap. 6 is valid and that the principal oscillator quantum number $N \leq 3$.

Problem 9.

1. Calculate the magnetic moment of nuclei with 3, 7 and 9 protons, for states with $m = j$. Disregard the neutron contribution.
2. Do the same for neutrons.

Problem 10. Repeat the calculation of Sect. 7.4.1 for a two-dimensional gas model.

Problem 11. The semiconductor Cu_2O displays an energy gap of 2.1 eV. If a thin sheet of this material is illuminated with white light:

1. What is the shortest wavelength that gets through?
2. What color is it?

Problem 12. Find the matrix elements $\langle ab|Q|ab\rangle$, $\langle bc|Q|ab\rangle$ and $\langle ac|Q|ab\rangle$ of the operator $\hat{Q} = \sum_{pq} q_{pq} a_p^+ a_q$, where a_p^+, a_p are fermion creation and annihilation operators and $p = a, b, c$. Assume that $q_{pq} = q_{qp}$.

8 Approximate Solutions to Quantum Problems

The previous chapters may have left the (erroneous) impression that there is always an exact (and elegant) mathematical solution for every problem in quantum mechanics. In most cases there is not. One must resort to makeshift approximations, numerical solutions or combinations of both. In this chapter we discuss three approximate methods that are frequently applied.

8.1 Perturbation Theory

The procedure is similar to the one used in celestial mechanics, where the trajectory of a comet is first calculated by taking into account only the attraction of the sun. The (smaller) effect of planets is included in successive orders of approximation.

We divide the Hamiltonian $\hat{H}$, which we do not know how to solve exactly, into two terms. The first term, $\hat{H}_0$, is the Hamiltonian of a problem whose solution we know and which is reasonably close to the original problem; the second term, $\hat{V}$, is called the perturbation. Thus

$$\hat{H} = \hat{H}_0 + \lambda \hat{V} , \tag{8.1}$$

$$\hat{H}_0 \varphi_n^{(0)} = E_n^{(0)} \varphi_n^{(0)} . \tag{8.2}$$

The perturbation term has been multiplied by a constant λ that is supposed to be a number less than 1. The constant λ is helpful for keeping track of the order of magnitude of the different terms of the expansion that underlies the theory. Otherwise it has no physical meaning. It is replaced by 1 in the final expressions. We solve the eigenvalue equation

$$\hat{H} \Psi_n = E_n \Psi_n \tag{8.3}$$

by expanding the eigenvalues and the eigenstates in powers of λ and successively considering all terms corresponding to the same power of λ in (8.3):

$$\begin{aligned} E_n &= E_n^{(0)} + \lambda E_n^{(1)} + \lambda^2 E_n^{(2)} + \cdots , \\ \Psi_n &= \varphi_n^{(0)} + \lambda \Psi_n^{(1)} + \lambda^2 \Psi_n^{(2)} + \cdots . \end{aligned} \tag{8.4}$$

The terms independent of λ yield (8.2). The terms proportional to λ give rise to the equation

$$\left(\hat{H}_0 - E_n^{(0)}\right)\Psi_n^{(1)} = \left(-\hat{V} + E_n^{(1)}\right)\varphi_n^{(0)}. \tag{8.5}$$

First, we take the scalar product of $\varphi_n^{(0)}$ with the states on each side of (8.5). The left-hand side vanishes because of (8.2). We thus obtain the first order correction to the energy

$$E_n^{(1)} = \langle \varphi_n^{(0)} | V | \varphi_n^{(0)} \rangle. \tag{8.6}$$

Therefore, the leading order term correcting the unperturbed energy is the expectation value of the perturbation.

Next, we take the scalar product with $\varphi_p^{(0)}$, $(p \neq n)$, so that

$$\left(E_p^{(0)} - E_n^{(0)}\right)\langle \varphi_p^{(0)} | \Psi_n^{(1)} \rangle = -\langle \varphi_p^{(0)} | V | \varphi_n^{(0)} \rangle. \tag{8.7}$$

Using the states $\varphi_p^{(0)}$ as basis states, we expand

$$\Psi_n^{(1)} = \sum_{p \neq n} c_p^{(1)} \varphi_p^{(0)}, \qquad c_p^{(1)} = \frac{\langle \varphi_p^{(0)} | V | \varphi_n^{(0)} \rangle}{E_n^{(0)} - E_p^{(0)}}. \tag{8.8}$$

The still missing amplitude $c_n^{(1)}$ is determined from the normalization condition: since both Ψ_n and $\varphi_n^{(0)}$ are supposed to be normalized to unity, the terms linear in λ are

$$0 = \langle \Psi_n | \Psi_n \rangle - \langle \varphi_n^{(0)} | \varphi_n^{(0)} \rangle = \lambda\left[\langle \Psi_n^{(1)} | \varphi_n^{(0)} \rangle + \langle \varphi_n^{(0)} | \Psi_n^{(1)} \rangle\right] = 2\lambda \mathrm{Re}\left(c_n^{(1)}\right). \tag{8.9}$$

Therefore, the first order coefficient $c_n^{(1)}$ disappears, since we can make it real by changing the (arbitrary) phase of $\varphi_n^{(0)}$.

Equations (8.6) and (8.8) determine the first order changes in the energies and state vectors in terms of matrix elements of the perturbation with respect to the basis of zero-order states. The convergence of perturbation theory requires that $|c_p^{(1)}|^2 \ll 1$, i.e., the matrix element of the perturbation between two states should be smaller than the unperturbed distance between these states. In particular, perturbation theory cannot be applied if there are non-vanishing matrix elements between degenerate states. In these cases, we must resort to either variational (Sect. 8.2) or diagonalization procedures (Sect. 8.5).

The second order correction to the energy is given by

$$E_n^{(2)} = \sum_{p \neq n} \frac{|\langle \varphi_p^{(0)} | V | \varphi_n^{(0)} \rangle|^2}{E_n^{(0)} - E_p^{(0)}}. \tag{8.10}$$

This perturbation theory is called the Raleigh–Schrödinger perturbation theory. Its apparent simplicity disappears in higher orders of perturbation, due to the increase in the number of contributing terms. A formal simplification may be achieved by summing up partial series of terms. For instance, in the Brillouin–Wigner perturbation theory, one replaces the unperturbed energy $E_n^{(0)}$ of the state n by the exact energy E_n in the denominators. For the case of the energy expansion, one obtains

$$
\begin{aligned}
E_n &= E_n^{(0)} + \langle \varphi_n^{(0)} | V | \varphi_n^{(0)} \rangle + \sum_{p \neq n} \frac{|\langle \varphi_p^{(0)} | V | \varphi_n^{(0)} \rangle|^2}{E_n - E_p^{(0)}} + \cdots \\
&= E_n^{(0)} + \langle \varphi_n^{(0)} | V | \varphi_n^{(0)} \rangle + \sum_{p \neq n} \frac{|\langle \varphi_p^{(0)} | V | \varphi_n^{(0)} \rangle|^2}{E_n^{(0)} - E_p^{(0)}} \\
&\quad - \sum_{p \neq n} \frac{|\langle \varphi_p^{(0)} | V | \varphi_n^{(0)} \rangle|^2 \langle \varphi_n^{(0)} | V | \varphi_n^{(0)} \rangle}{\left(E_n^{(0)} - E_p^{(0)}\right)^2} + \cdots .
\end{aligned}
\tag{8.11}
$$

The last term appears as a third order term in the Raleigh–Schrödinger perturbation theory. It does not exist[1] in the Brillouin–Wigner expansion, since it has been taken into account through the replacement in the denominator of the second order term (8.10). However, the advantage of reducing the number of terms may be compensated by a decrease in the convergence of the perturbation expansion, associated with the nature of the partial summations. Moreover, different powers of λ may be present in many terms of the Brillouin–Wigner series.

There is an elegant and useful formulation of perturbation theory conceived by Feynman. This uses diagrams carrying both a precise mathematical meaning and a description of the processes involved [52]. The 'finest hour' of perturbation theory is represented by the calculation of the Lamb shift, i.e., the energy difference $E_{2p\frac{1}{2}} - E_{1s\frac{1}{2}}$ in the hydrogen atom, to six significant figures, using quantum electrodynamics (see [53], p. 358).

The ground state energy of the He atom is calculated using perturbation theory in Sect. 8.3.

8.2 Variational Procedure

This approximation may be considered as the inverse of the perturbation procedure: instead of working with a fixed set of unperturbed states, one guesses a trial state Ψ, which may be expanded in terms of the basis set of eigenstates φ_E of the Hamiltonian ($\Psi = \sum_E c_E \varphi_E$):

---

[1] One can prove that the Brillouin–Wigner expansion does not contain terms in which the state $\varphi_n^{(0)}$ appears in the numerator as an intermediate state.

$$\langle\Psi|H|\Psi\rangle = \sum_E E|c_E|^2 \geq E_0 \sum_E |c_E|^2 = E_0 , \qquad (8.12)$$

where E_0 is the ground state energy. The state Ψ may depend on some parameter, and the expectation value of the Hamiltonian is minimized with respect to this parameter. One thus obtains an upper limit for the ground state energy of the system.

The fact that the energy is an extremum guarantees that if the trial wave function is wrong by something of the order of δ, the variational estimate of the energy is off by something of the order δ^2. So one can be rewarded with a very good energy estimate, even though the initial wave function is only a fair guess.

The ground state energy obtained in first order perturbation theory $E_0^{(0)} + \langle\varphi_0^{(0)}|V|\varphi_0^{(0)}\rangle$ is an expectation value of the total Hamiltonian, and is thus equivalent to a non-optimized variational calculation. An application of the variational procedure to the case of the He atom can also be found in Sect. 8.3.

8.3 Ground State of the He Atom

This three-body problem may be reduced to a two-body problem by again considering a very massive nucleus. However, even the remaining problem is difficult to solve because of the presence of the Coulomb repulsion V between the two electrons. The total Hamiltonian is

$$H_0 = -\frac{\hbar^2}{2M}\left(\nabla_1^2 + \nabla_2^2\right) - \frac{Ze^2}{4\pi\epsilon_0}\left(\frac{1}{r_1} + \frac{1}{r_2}\right), \qquad V = \frac{e^2}{4\pi\epsilon_0 r_{12}} . \qquad (8.13)$$

We know how to solve the problem of two electrons moving independently of each other in the Coulomb potential of the He nucleus. Because the ground state energy of a hydrogen-like atom is proportional to Z^2 and there are two electrons, the unperturbed energy is $8E_H$, where E_H is the energy of the electron in the H atom. The antisymmetrized two-electron state vector of the ground state in the He atom is discussed in Sect. 7.2. Using this state, the first order correction to the energy is (Sect. 8.6*)

$$E_{\text{g.s.}}^{(1)} = \left\langle\varphi_{\text{g.s.}}\left|\frac{e^2}{4\pi\epsilon_0 r_{12}}\right|\varphi_{\text{g.s.}}\right\rangle = -\frac{5}{2}E_H . \qquad (8.14)$$

Therefore, the total energy becomes $5.50 E_H$, which constitutes an improved approximation to the experimental result $5.81 E_H$ compared with the unperturbed value $8E_H$.

As mentioned in Sect. 8.2, one may improve the predictions for the ground state by a variational calculation. In this case we may write the expectation value of the kinetic energy, of the potential energy and the Coulomb repulsion as functions of the parameter Z^* entering the wave function. The value $Z = 2$ is kept in the Hamiltonian:

$$\langle \varphi_{\text{g.s.}} | H | \varphi_{\text{g.s.}} \rangle_{Z^*} = -2(Z^*)^2 E_\text{H} + 4ZZ^* E_\text{H} - \frac{5}{4} Z^* E_\text{H} \, . \tag{8.15}$$

Minimization with respect to Z^* yields the effective value $Z^* = 1.69$ for He (instead of 2), which is an indication that the electrons mutually screen the nuclear attraction. The final result is $\langle \varphi_{\text{g.s.}} | H | \varphi_{\text{g.s.}} \rangle_{Z=1.69} = 5.69 E_\text{H}$, and this is in even better agreement with the experimental value than the first order perturbation result.

In order to apply the variational procedure to excited states, one must ensure their orthogonality with lower energy states, for the resulting value of the minimization parameter.

8.4 Molecules

Molecules are made up of nuclei and electrons. The theoretical description of this many-body system is facilitated by the very different masses of the two constituents, which allows one to decouple their respective motions. The procedure is called the Born–Oppenheimer approximation. In principle, it is possible to begin by solving the problem of motion of electrons subject to the (static) field of the nuclei and to the field of other electrons. In this first step, the nuclear coordinates $\boldsymbol{R}_i$ are treated as parameters. Minimization of the energy $W(\boldsymbol{R}_i)$ with respect to these parameters yields their equilibrium values. A subsequent step consists in allowing small departures of the nuclei from their equilibrium position and using the associated increase in the energy W as the restoring force for the oscillatory motion. Finally, the molecules may also perform collective rotations without changing the relative positions of the electrons and the nuclei.

8.4.1 Intrinsic Motion. Covalent Binding

We illustrate the procedure for the case of the molecular hydrogen ion H_2^+. Figure 8.1 represents the two protons 1 and 2 and the electron. The assumption that the protons are at rest simplifies the Hamiltonian to

$$\hat{H} = -\frac{\hbar^2}{2M} \nabla^2 - \frac{e^2}{4\pi\epsilon_0 |\boldsymbol{r} - \boldsymbol{R}_1|} - \frac{e^2}{4\pi\epsilon_0 |\boldsymbol{r} - \boldsymbol{R}_2|} + \frac{e^2}{4\pi\epsilon_0 R} \, , \tag{8.16}$$

where $R = |\boldsymbol{R}_1 - \boldsymbol{R}_2|$ is the distance between the protons. Although in this particular case exact numerical solutions may be obtained by solving the Schrödinger equation in elliptical coordinates, it is more instructive to approximate the solution by means of a variational procedure.

8 Approximate Solutions to Quantum Problems

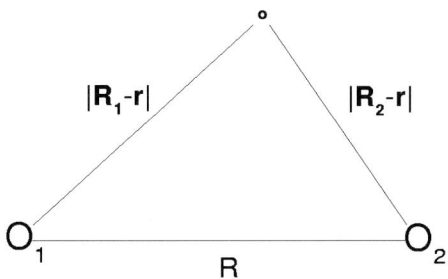

Fig. 8.1. The hydrogen ion

If the distance R is very large, the two (degenerate) solutions describe a H atom plus a dissociated proton. The two orbital wave functions are

$$\varphi_1 = \varphi_{100}(|\boldsymbol{r} - \boldsymbol{R}_1|), \qquad \varphi_2 = \varphi_{100}(|\boldsymbol{r} - \boldsymbol{R}_2|). \tag{8.17}$$

Note that such wave functions are orthogonal only for very large values of R. In fact, their overlap is $\langle 1|2\rangle = 1$ for $R = 0$.

The requirement of antisymmetry between the two protons must be taken into account. As in Sect. 7.2, the spin of the two protons may be coupled to 1 (symmetric spin states) or to 0 (antisymmetric spin states). The corresponding spatial wave functions should thus be antisymmetric or symmetric, respectively:

$$\varphi_\mp = \frac{\varphi_1 \mp \varphi_2}{\sqrt{2(1 \mp \langle 1|2\rangle)}}. \tag{8.18}$$

The energy to be minimized with respect to the distance R is

$$\begin{aligned}W_\pm(R) &= \langle \pm|H|\pm\rangle \\ &= E_{100} + \frac{e^2}{4\pi\epsilon_0 R} - \frac{e^2}{4\pi\epsilon_0(1 \pm \langle 1|2\rangle)}\left\langle 2\left|\frac{1}{|\boldsymbol{r}-\boldsymbol{R}_1|}\right|2\right\rangle \\ &\quad \mp \frac{e^2}{4\pi\epsilon_0(1 \pm \langle 1|2\rangle)}\left\langle 1\left|\frac{1}{|\boldsymbol{r}-\boldsymbol{R}_1|}\right|2\right\rangle, \end{aligned} \tag{8.19}$$

which has the limits

$$\lim_{R\to 0} W \to \infty, \qquad \lim_{R\to\infty} = E_{100}. \tag{8.20}$$

Since the matrix element in the third line of (8.19) is positive, we conclude that the energy corresponding to the spatially symmetric wave function lies lowest. In fact, the two curves are plotted in Fig. 8.2. Only the energy corresponding to φ_+ displays a minimum. This may be interpreted as being due to the buildup of the electron density between the two nuclei, which allows for the screening of the Coulomb repulsion. This type of binding is called covalent binding.

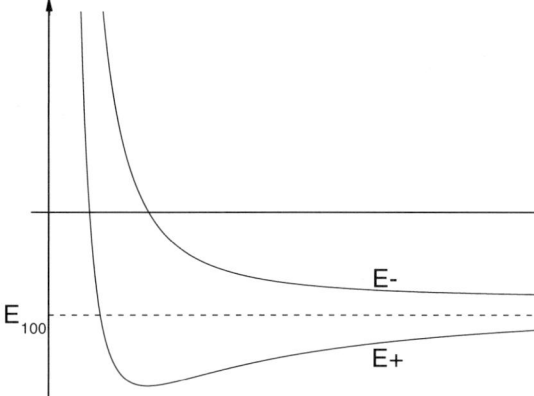

Fig. 8.2. Lowest energies of the hydrogen ion as a function of the distance between protons

For future comparisons, let us state that the characteristic intrinsic excitation energies of electrons in molecules are of the same order of magnitude as in atoms, since the Coulomb interaction between electrons and nuclei keeps the molecules together. Thus,

$$E_{\text{intr}} \approx -E_{100} \approx \frac{e^2}{8\pi\epsilon_0 a_0} = \frac{\hbar^2}{Ma_0^2}, \qquad (8.21)$$

where a_0 is the Bohr radius (Table 6.1).

8.4.2 Vibrational and Rotational Motions

We consider here the somewhat more general case of a diatomic molecule with masses M_1 and M_2, respectively. First we perform the well known separation between the relative and center of mass operators:

$$\hat{\boldsymbol{R}} = \hat{\boldsymbol{R}}_1 - \hat{\boldsymbol{R}}_2, \qquad \hat{\boldsymbol{R}}_g = \frac{M_1}{M_g}\hat{\boldsymbol{R}}_1 + \frac{M_2}{M_g}\hat{\boldsymbol{R}}_2,$$

$$\hat{\boldsymbol{P}} = \frac{M_2}{M_g}\hat{\boldsymbol{P}}_1 - \frac{M_1}{M_g}\hat{\boldsymbol{P}}_2, \qquad \hat{\boldsymbol{P}}_g = \hat{\boldsymbol{P}}_1 + \hat{\boldsymbol{P}}_2. \qquad (8.22)$$

The inversion of definitions (8.22) yields the kinetic energy

$$\frac{\hat{\boldsymbol{P}}_1^2}{2M_1} + \frac{\hat{\boldsymbol{P}}_2^2}{2M_2} = \frac{\hat{\boldsymbol{P}}_g^2}{2M_g} + \frac{\hat{\boldsymbol{P}}^2}{2\mu}. \qquad (8.23)$$

Here $M_g = M_1 + M_2$ is the total mass and $\mu \equiv M_1 M_2/M_g$ is the reduced mass.

8 Approximate Solutions to Quantum Problems

If the potential energy $V(R)$ depends only on the distance between the ions, the center of mass moves as a free particle. This problem has already been discussed in Sect. 4.3. The kinetic energy associated with the relative motion may be expressed in spherical coordinates, as in (6.1), with the substitution $M \to \mu$.

Let us split the relative Hamiltonian into rotational and vibrational contributions, viz.,

$$\hat{H} = \hat{H}_{\mathrm{rot}} + \hat{H}_{\mathrm{vib}} \, ,$$

$$\hat{H}_{\mathrm{rot}} = \frac{1}{2\mu R^2} \hat{\boldsymbol{L}}^2 \, , \qquad (8.24)$$

$$\hat{H}_{\mathrm{vib}} = -\frac{\hbar^2}{2\mu} \left(\frac{\mathrm{d}^2}{\mathrm{d}R^2} + \frac{2}{R} \frac{\mathrm{d}}{\mathrm{d}R} \right) + V(R) \, . \qquad (8.25)$$

We now assume that the interactions between the ions stabilize the system at the relative distance R_0. The difference $y = R - R_0$ will be such that $|y| \ll R_0$.

Rotational Motion

The Hamiltonian for the rotational motion may be approximated as

$$\hat{H}_{\mathrm{rot}} = \frac{1}{2\mu R^2} \hat{\boldsymbol{L}}^2 \approx \frac{1}{2\mu R_0^2} \hat{\boldsymbol{L}}^2 \, . \qquad (8.26)$$

The eigenfunctions are labeled by the quantum numbers l, m_l (Sect. 5.1.2). The energies are obtained by replacing the operator $\hat{\boldsymbol{L}}^2$ in (8.26) by its eigenvalues $\hbar^2 l(l+1)$. The photon energy corresponding to the transition between neighboring states increases linearly with l, so that

$$\Delta(l \to l-1) = \frac{\hbar^2}{\mu R_0^2} l \, . \qquad (8.27)$$

The rotational energies are of the order of

$$E_{\mathrm{rot}} \approx \frac{\hbar^2}{\mu a_0^2} \approx \frac{M}{M_{\mathrm{p}}} E_{\mathrm{intr}} \, , \qquad (8.28)$$

where we have replaced the reduced mass μ by the proton mass M_{p}. Since the ratio $M/M_{\mathrm{p}} = O(1/2000)$, we expect different orders of magnitude for the frequency of the radiation emitted (or absorbed) in transitions between intrinsic and rotational excitations.

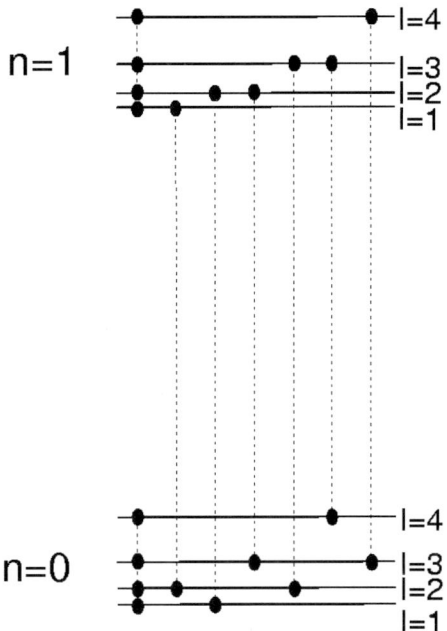

Fig. 8.3. Vibrational and rotational excitations of a molecule. *Dotted lines* represent allowed transitions, according to the definition (9.60)

Vibrational Motion

If the stabilization at $R \approx R_0$ is sufficiently good we may extend the domain of the radial coordinate from 0 to $-\infty$, since the wave function should be increasingly small for negative values of R. Simultaneously, the R^2 factor in the volume element may be eliminated from the integrals by the substitution $\Psi(R) \to \Phi(R)/R$. In such a case the radial Schrödinger equation transforms into a linear equation of the type seen in Chap. 4:

$$-\frac{\hbar^2}{2\mu}\left(\frac{d^2}{dR^2} + \frac{2}{R}\frac{d}{dR}\right)\Psi + V(R)\Psi = E\Psi \longrightarrow -\frac{\hbar^2}{2\mu}\frac{d^2}{dR^2}\Phi + V(R)\Phi = E\Phi, \tag{8.29}$$

with the boundary conditions $\Phi(\pm\infty) = 0$.

Finally, the Taylor expansion of the potential around the equilibrium position R_0, and the replacement of the coordinate R by $y = R - R_0$, yield the harmonic oscillator Hamiltonian discussed in Sects. 3.2 and 4.2:

$$\left[-\frac{\hbar^2}{2\mu}\frac{d^2}{dy^2} + \frac{1}{2}\left.\frac{d^2V(R)}{dR^2}\right|_{R=R_0} y^2\right]\Phi = \left[E - V(R_0)\right]\Phi. \tag{8.30}$$

The vibrational states are equidistant from each other (Fig. 3.2). The photon spectrum displays the single frequency

$$\Delta(N \to N-1) = \hbar\omega = \hbar\sqrt{\frac{1}{\mu}\frac{\mathrm{d}^2V(R)}{\mathrm{d}R^2}\bigg|_{R=R_0}}. \tag{8.31}$$

The vibrational energies are of the order of

$$E_{\text{vib}} \approx \hbar\sqrt{\frac{E_{\text{intr}}}{a_0^2\mu}} \approx \frac{\hbar^2}{a_0^2\sqrt{M\mu}} \approx \sqrt{\frac{M}{M_{\text{p}}}} E_{\text{intr}}, \tag{8.32}$$

i.e., the photons corresponding to the transitions between vibrational states occupy an intermediate energy range compared to those corresponding to intrinsic electron transitions or to transitions between rotational states (Fig. 8.3).

As the energies of the rotational and vibrational excitations increase, the approximations become less reliable:

- terms that are functions of y will appear in the rotational Hamiltonian, coupling the rotational and vibrational motion;
- higher order terms in the Taylor expansion of the potential become relevant.

8.5 Approximate Matrix Diagonalizations

If the conditions for applying perturbation theory are not satisfied, we may resort to a diagonalization procedure. This is obviously necessary if there are degenerate or close-lying states. This is the case if two or more particles are added to a closed shell, whether it be atomic or nuclear. The size of the matrix to be diagonalized may be reduced due to physical considerations, for example, when we use the symmetries of the Hamiltonian. If we are only interested in the ground state and neighboring states, we may also simplify the problem by taking into account only those states which are close in energy to the ground state.

It is also possible to include those contributions to the matrix elements of the Hamiltonian to be diagonalized that arise from states not included in the diagonalization. One may use either the technique of folded diagrams (a generalization of Raleigh–Schrödinger perturbation theory) [56] or the Bloch–Horowitz procedure (an extension of the Brillouin–Wigner expansion) [57].

An alternative procedure consists in simplifying the expressions for the matrix elements. This includes eliminating many of them (see Problem 12). In such cases, good insight is required in order to avoid distorting the physical problem.

8.5.1* Approximate Treatment of Periodic Potentials

This example illustrates the interplay between exact diagonalization and perturbation theory that can be applied in more complicated situations. We treat

8.5 Approximate Matrix Diagonalizations

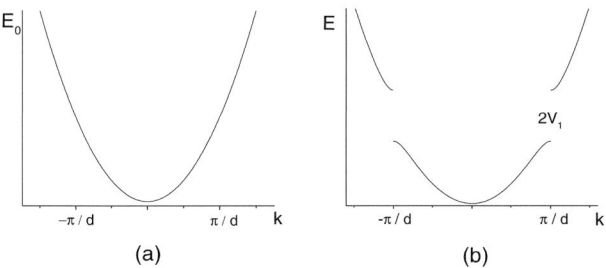

Fig. 8.4. Bands in periodic potentials. The unperturbed parabolic energies are given as a function of k in (a). The eigenvalue E_- is plotted in the interval $0 \leq |k| \leq \frac{\pi}{d}$ and E_+ for $\frac{\pi}{d} \leq |k|$ in (b)

the same problem as in Sect. 4.7*, but in the limit of a small periodic potential $V(x)$.

We choose the free-particle Hamiltonian $H_0 = \frac{1}{2M}\hat{p}^2$ as zero-order Hamiltonian. The unperturbed energies are given in Fig. 8.4a, as a function of the wave number k. If $V(x) = V(x+d)$, a Fourier transform of the potential yields

$$V(x) = \sum_n V_n \; ; \quad V_n = v_n \exp\left(\frac{i2\pi n x}{d}\right) \; ; \quad n = 0, \pm 1, \pm 2, \dots \quad (8.33)$$

Therefore the non-vanishing matrix elements of the perturbation are

$$\langle k'|V_n|k\rangle = v_n \quad \text{if } k' - k = \frac{2\pi n}{d}, \quad (8.34)$$

and the Hamiltonian matrix is of the form[2]

$$\begin{pmatrix} \dots & \dots & \dots & \dots & \dots & \dots \\ \dots & \frac{\hbar^2}{2M}\left(k - \frac{4\pi}{d}\right)^2 & v_1 & v_2 & v_3 & \dots \\ \dots & v_1 & \frac{\hbar^2}{2M}\left(k - \frac{2\pi}{d}\right)^2 & v_1 & v_2 & \dots \\ \dots & v_2 & v_1 & \frac{\hbar^2}{2M}k^2 & v_1 & \dots \\ \dots & v_3 & v_2 & v_1 & \frac{\hbar^2}{2M}\left(k + \frac{2\pi}{d}\right)^2 & \dots \\ \dots & \dots & \dots & \dots & \dots & \dots \end{pmatrix} \quad (8.35)$$

We concentrate on the non-diagonal terms v_1. However small, they cannot be treated in perturbation theory, because they connect degenerate states: the state with $k = \frac{\pi}{d}$ has the same unperturbed energy as the state with $k' = k - \frac{2\pi}{d} = -\frac{\pi}{d}$. Thus, we must first proceed to make a diagonalization between degenerate (or quasi-degenerate) states, i.e., we must put to zero the determinant

---
[2] We disregard v_0 since only affects the zero-point energy

$$\begin{vmatrix} \frac{\hbar^2}{2M}\left(k-\frac{2\pi}{d}\right)^2 - E & v_1 \\ v_1 & \frac{\hbar^2}{2M}k^2 - E \end{vmatrix} = 0 \qquad (8.36)$$

The eigenvalues

$$E_\pm = \frac{\hbar^2}{2M}\left(k^2 - k\frac{2\pi}{d} + \frac{2\pi^2}{d^2} \pm \sqrt{\frac{4\pi^2}{d^2}\left(k-\frac{\pi}{d}\right)^2 + \left(\frac{2Mv_1}{\hbar^2}\right)^2}\right) \qquad (8.37)$$

are plotted as a function of k in Fig. 8.4b. There are no states in the interval $\frac{1}{2M}\left(\frac{\hbar\pi}{d}\right)^2 - v_1 \leq E \leq \frac{1}{2M}\left(\frac{\hbar\pi}{d}\right)^2 + v_1$. A gap of size $2v_1$ appears in the spectrum, pointing to the existence of two separate bands.

In the region $|k| \approx \frac{\pi}{d}$, the remaining non-diagonal terms v_n may be treated as a perturbation, if they are sufficiently small. Unfortunately, they usually are not so in realistic cases.

8.6* Matrix Elements Involving the Inverse of the Interparticle Distance

Although the integrals involved may be found in tables, we calculate them explicitly as a quantum mechanical exercise. The inverse of the distance between two particles may be expanded as

$$\frac{1}{r_{12}} = \frac{1}{r_2}\sum_l \left(\frac{r_1}{r_2}\right)^l P_l(\cos\alpha_{12}), \qquad r_1 < r_2. \qquad (8.38)$$

Here P_l is the Legendre polynomial of order l (Sect. 5.5*), a function of the angle α_{12} subtended by the two vectors $\mathbf{r}_1, \mathbf{r}_2$. It may be expressed by coupling two spherical harmonics to zero angular momentum (5.51).

Next, we evaluate matrix elements such as

$$\left\langle n_1 l_1 m_1 n_2 l_2 m_2 \left| \frac{1}{r_{12}} \right| n_1 l_1 m_1 n_2 l_2 m_2 \right\rangle \qquad (8.39)$$

$$= N^2_{n_1 l_1} N^2_{n_2 l_2} \int_0^\infty |R_{n_1 l_1}(1)|^2 r_1^2 dr_1 \int_0^\infty |R_{n_2 l_2}(2)|^2 r_2^2 dr_2$$

$$\times \int_0^{4\pi} |Y_{l_1 m_1}(1)|^2 d\Omega_1 \int_0^{4\pi} |Y_{l_2 m_2}(2)|^2 d\Omega_2 / r_{12}$$

$$= N^2_{n_1 l_1} N^2_{n_2 l_2} \sum_l \frac{4\pi}{2l+1} \int_0^\infty |R_{n_1 l_1}(1)|^2 r_1^2 dr_1$$

$$\times \left[\frac{1}{r_1^{l+1}} \int_0^{r_1} |R_{n_2 l_2}(2)|^2 r_2^{l+2} dr_2 + r_1^l \int_{r_1}^\infty |R_{n_2 l_2}(2)|^2 r_2^{1-l} dr_2 \right]$$

$$\times \sum_{m_l=-l}^{m_l=l} (-1)^{l-m_l} \langle Y_{l_1 m_1}|Y_{l m_l}|Y_{l_1 m_1}\rangle \langle Y_{l_2 m_2}|Y_{l(-m_l)}|Y_{l_2 m_2}\rangle .$$

The angular integrals restrict the values of l in the summation (see Problem 5 in Chap. 5). If at least one of the particles is in an s state, only one l term survives. If both particles are in s states,

$$\left\langle n_1 0 0 n_2 l_2 m_2 \left| \frac{1}{r_{12}} \right| n_1 0 0 n_2 l_2 m_2 \right\rangle = N_{n_1 0}^2 N_{n_2 l_2}^2 \int_0^\infty |R_{n_1 0}(1)|^2 r_1^2 \mathrm{d}r_1 \quad (8.40)$$
$$\times \left[\frac{1}{r_1} \int_0^{r_1} |R_{n_2 l_2}(2)|^2 r_2^2 \mathrm{d}r_2 + \int_{r_1}^\infty |R_{n_2 l_2}(2)|^2 r_2 \mathrm{d}r_2 \right],$$

which yields the value (8.14) for $n_1 = n_2 = 1$.

8.7* Quantization with Constraints

Little attention, if any, is paid in quantum textbooks to the problem of quantization with constraints, an area where great progress has been made over the last 30 years [58, 59]. This subject is not only of paramount importance in gauge field theories, but it also has applications in quantum mechanics, as in the description of many-body systems in moving frames of reference [60]. Moreover, the problem is conceptually significant in terms of properties of Hilbert spaces.

In this presentation we use two-dimensional rotations to exemplify a description made in terms of an overcomplete set of degrees of freedom. This overcompleteness requires the existence of constraints.

8.7.1* Constraints

We shall call intrinsic the coordinates of a system that refer to a rotating frame of reference. The motion of the moving frame relative to the laboratory is described by means of collective coordinates. Therefore, in this problem:

- There is an overcomplete set of angular variables (intrinsic + collective) describing transformations (rotations) of the system.
- The rotations of the system are generated (5.10) by the intrinsic angular momentum $\hat{L}$. There is also a collective angular momentum $\hat{I}$, the generator of rotations of the moving frame.
- The classical set of equations defining the momenta in terms of partial derivatives of the Lagrange function $\mathcal{L}$ cannot be solved in this case. This failure is due to the fact that this function does not contain information about the frame itself. For instance, in the case of one particle allowed to move on a circumference of radius r_0, the Lagrange function may be expressed in terms of the angular velocities $\dot{\alpha}$ and $\dot{\phi}$ (Fig. 8.5):

$$\mathcal{L} = \frac{\mathcal{J}}{2} \left(\dot{\alpha} + \dot{\phi} \right)^2 . \quad (8.41)$$

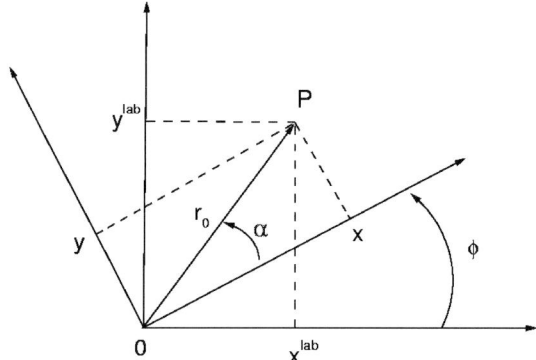

Fig. 8.5. Intrinsic (x, y) and laboratory $(x^{\text{lab}}, y^{\text{lab}})$ coordinates of a generic point P. The two sets of coordinates are related by a rotation [61]

Here $\alpha = \tan^{-1}(y/x)$ and $\mathcal{J} = Mr_0^2$ is the moment of inertia. From the equations

$$L = \frac{\partial \mathcal{L}}{\partial \dot\alpha} \quad \text{and} \quad I = \frac{\partial \mathcal{L}}{\partial \dot\phi},$$

one obtains the orbital angular momentum L and the constraint $f = 0$:

$$L = \mathcal{J}\left(\dot\alpha + \dot\phi\right), \tag{8.42}$$

$$f \equiv L - I = 0. \tag{8.43}$$

Equation (8.43) expresses the obvious fact that if the particle is rotated through an angle relative to the moving frame, the corresponding description should be completely equivalent to the one obtained by rotating the moving frame in the opposite direction. This constitutes a mechanical analogue of gauge invariance.

- Our aim is to quantize this classical model. The following commutation relations hold:

$$[\hat\alpha, \hat L] = [\hat\phi, \hat I] = i\hbar. \tag{8.44}$$

Since we have artificially enlarged the vector space, we must expect the presence of unphysical states and operators, in addition to physical ones. The constraint (8.43) is equivalent to the quantum mechanical conditions

$$\hat f \varphi_{\text{ph}} = 0, \quad \hat f \varphi_{\text{unph}} \neq 0,$$
$$[\hat f, \hat O_{\text{ph}}] = 0, \quad [\hat f, \hat O_{\text{unph}}] \neq 0. \tag{8.45}$$

where the labels ph and unph indicate physical and unphysical states or operators. Except in simple cases, this separation is by no means a trivial operation.

- Since the problem displays cylindrical symmetry, there is no restoring force in the intrinsic angular direction and perturbation theory cannot be applied.

8.7.2* Outline of the BRST Solution

The most natural thing to do would be to use the constraint in order to reduce the number of variables to the initial number. However, progress has been made in the opposite direction, i.e., by enlarging the number of variables and introducing a more powerful symmetry.

The collective subspace is given by the eigenfunctions of the orbital angular momentum in two dimensions (5.46):

$$\varphi_m(\phi) = \frac{1}{\sqrt{2\pi}} \exp(im\phi) . \tag{8.46}$$

The collective coordinate ϕ, which was introduced in Sect. 8.7.1* as an artifact associated with the existence of the moving frame, has been raised to the status of a real degree of freedom.

Since this problem has only one real degree of freedom, and since this role is taken by the collective angle, all others are unphysical. There must therefore be a trade-off: the intrinsic coordinate α has to be transferred to the unphysical subspace. In the Becchi–Rouet–Stora–Tyutin procedure, this subspace is also integrated with auxiliary fields [58]. All effects of the unphysical degrees of freedom on any physical observable cancel out. Moreover, the degree of freedom α acquires a finite frequency through its mixture with the other spurious fields, and perturbation theory becomes feasible.

Unfortunately, we cannot go beyond this point here. An elementary presentation of the quantum mechanical BRST method is given in [61].

Problems

Problem 1.

1. Obtain the expression for the second order correction to the energy in perturbation theory and show that this correction is always negative for the ground state.
2. Calculate the second order correction to the eigenstate.

8 Approximate Solutions to Quantum Problems

Problem 2. Assume that the zero order Hamiltonian and the perturbation are given by the matrices

$$\hat{H}_0 = \begin{pmatrix} 5 & 0 & 0 \\ 0 & 2 & 0 \\ 0 & 0 & -1 \end{pmatrix}, \quad \hat{V} = \begin{pmatrix} 0 & c & 0 \\ c & 0 & 0 \\ 0 & 0 & 2c \end{pmatrix}.$$

1. Calculate the first order perturbation corrections to the energies.
2. Calculate the second order perturbation corrections to the energies.
3. Obtain the first order corrections to the vector states.
4. Obtain the second order corrections to the vector states.
5. Expand the exact energies in powers of c and compare the results with those obtained in perturbation theory:

$$(1+x)^{1/2} = 1 + \frac{1}{2}x - \frac{1}{8}x^2 + \frac{1}{16}x^3 - \cdots.$$

Problem 3.

1. Calculate the first and second order corrections to the ground state energy of a linear harmonic oscillator if a perturbation $V(x) = kx$ is added, and compare with the exact value.
2. Do the same if the perturbation is $V(x) = bx^2/2$.

Problem 4. Substitute $R \to R_0 + y$ in the rotational Hamiltonian (8.24) and expand the Hamiltonian in powers of y up to quadratic order. Calculate the correction for the energy in perturbation theory, using the product of the rotational and vibrational bases as an unperturbed basis:

$$(1+a)^{-2} = 1 - 2a + 3a^2 + \cdots.$$

Problem 5.

1. Calculate the lowest relativistic correction to the ground state energy of a linear harmonic oscillator. Hint: expand the relativistic energy $\sqrt{M^2 c^4 + c^2 p^2}$ in powers of p/Mc.
2. Obtain the order of magnitude of the ratio between the relativistic correction and the non-relativistic value in the molecular case.

Problem 6. Obtain the vector state up to second order in the Brillouin–Wigner perturbation theory. Compare with the results (8.8) and Problem 1.

Problem 7. Show that the Brillouin–Wigner perturbation theory already yields the exact results (3.23) in second order, for a Hamiltonian of the form (3.22). Hint: use
$$E_a = \langle a|H|a\rangle + \frac{|\langle a|H|b\rangle|^2}{E_a - \langle b|H|b\rangle} .$$

Problem 8. Minimize the ground state energy by taking the mass as the variation in the lowest harmonic oscillator state, and using the harmonic oscillator potential plus the relativistic kinetic energy as Hamiltonian. Include as many powers of p^2/M^2c^2 in the latter as are necessary in order to obtain an improvement over the perturbation results of Problem 5.

1. Write the expectation value of the Hamiltonian as a function of M^*/M.
2. Write the minimization condition.
3. Solve this equation in powers of $\hbar\omega/Mc^2$.
4. Expand the energies in powers of $\hbar\omega/Mc^2$.

Problem 9. Calculate the perturbation correction for the two $1s2p$ electron states in the He atom. Explain why perturbation theory may be used in spite of the existing degeneracies.

Problem 10.

1. In units of E_H, calculate the first order perturbative correction for the ground state energy of the He atom, the ionized Li atom and the doubly ionized Be atom.
2. Obtain the variational energies using the effective number of electrons Z^* as the variational parameter.
3. Compare with the experimental values: -79 eV (He), -197 eV (Li$^+$), -370 eV (Be^{++}).

Problem 11. A hydrogen atom is subject to a constant electric field in the z-direction (Stark effect).

1. Construct the matrix of the perturbation for the $n = 2$ state and diagonalize this matrix.
2. Do the same for the $N = 2$ states of the harmonic oscillator potential.

Problem 12. Consider two fermions moving in a j-shell.

1. Calculate the size of the Hamiltonian matrix if we assume two-particle states of the form
$$\phi_{mm'} = \frac{1}{\sqrt{2}} \left[\varphi_{jm}(1)\varphi_{jm'}(2) - \varphi_{jm}(2)\varphi_{jm'}(1) \right] .$$

2. Approximate the matrix elements of the Hamiltonian by the expression $\langle mm'|H|m''m'''\rangle = -g\,\delta_{m(-m')}\delta_{m''(-m''')}$. Calculate the size of the (reduced) matrix to be diagonalized.
3. Find the eigenvectors and eigenvalues. Hint: Try a solution of the form $\Psi = \sum_m c_m \phi_{m(-m)}$: (a) with amplitudes $c_m = $ constant, (b) with amplitudes such that $\sum_m c_m = 0$.

Note that the particles in the resulting extra-bound state are said to form Cooper pairs. This extra binding is the basis for the explanation of the phenomenon of superconductivity [54].

9 Time-Dependence in Quantum Mechanics

9.1 The Time Principle

Up to now we have only considered static situations (except for the reduction of the state vector when a measurement takes place). We will now discuss time-dependence of the state vector. Assume that the system is represented at time t by the time-dependent state vector $\Psi(t)$. At time $t' > t$, the system will evolve in accordance with

$$\Psi(t') = \hat{U}(t',t)\Psi(t) , \qquad (9.1)$$

where $\hat{U}(t',t)$ is called the evolution operator. This operator satisfies the conditions of being unitary and

$$\lim_{t' \to t} U(t',t) = 1 . \qquad (9.2)$$

Therefore, if $t' = t + \Delta t$,

$$\Psi(t + \Delta t) = U(t + \Delta t, t)\Psi(t)$$
$$= \left[1 + \frac{\partial}{\partial t'}U(t',t)\bigg|_{t'=t} \Delta t + \cdots \right]\Psi(t) ,$$

$$\frac{\partial}{\partial t}\Psi(t) = \frac{\partial}{\partial t'}U(t',t)|_{t'=t}\Psi(t) . \qquad (9.3)$$

A new quantum principle must be added to those stated in Chaps. 2 and 7.

Principle 5. *The operator yielding the change of the state vector over time is proportional to the Hamiltonian*

$$\frac{\partial}{\partial t'}U(t',t)|_{t'=t} = -\frac{i}{\hbar}\hat{H}(t) . \qquad (9.4)$$

The time evolution of the system is determined by the first order linear equation

$$i\hbar\frac{\partial}{\partial t}\Psi(t) = \hat{H}\Psi(t) . \qquad (9.5)$$

9 Time-Dependence in Quantum Mechanics

This is called the time-dependent Schrödinger equation. It is valid for a general state vector, and it is independent of any particular realization of quantum mechanics.

If $[\hat{H}(\tau_1), \hat{H}(\tau_2)] = 0$, the evolution operator is

$$\hat{U}(t', t) = \exp\left[i \int_t^{t'} \hat{H}(\tau) d\tau\right] . \tag{9.6}$$

In the case of a time-independent Hamiltonian satisfying the eigenvalue equation $H_0 \varphi_i = E_i \varphi_i$, the solution to the differential equation (9.5) may be found using the method of separation of variables. Hence,

$$\varphi_i(t) = f(t)\varphi_i , \qquad i\hbar \frac{df}{dt} = E_i f \longrightarrow f = \exp(-iE_i t/\hbar) \tag{9.7}$$

and thus,

$$\varphi_i(t) = \varphi_i \exp(-iE_i t/\hbar) . \tag{9.8}$$

We expect the solutions of a time-independent Hamiltonian to be independent of time. However, as usual, this requirement can only be enforced up to a phase. This is consistent with the result (9.8).

In particular, the time-dependent wave function for a free particle with energy $E = \hbar\omega$ is as expected (4.30):

$$\varphi_{\pm k}(x, t) = A \exp\left[i(\pm kx - \omega t)\right] . \tag{9.9}$$

If the Hamiltonian is time-independent and the state is represented at time $t = 0$ by the linear combination (2.3) of its eigenstates,

$$\Psi(t=0) = \sum_i c_i \varphi_i , \tag{9.10}$$

equation (9.8) implies that at time t the state has evolved into[1]

$$\Psi(t) = \sum_i c_i \varphi_i \exp(-iE_i t/\hbar) . \tag{9.11}$$

9.2 Time-Dependence of Spin States

9.2.1 Larmor Precession

To begin with we give a simple but non-trivial example of a solution to (9.5). The time-dependent equation associated with the interaction between

---

[1] This evolution is valid only for the Hamiltonian basis. Therefore, the expression $\Psi(t) = \sum_i c_i \phi_i \exp(-iq_i t/\hbar)$ makes no sense if ϕ_i, q_i are not eigenstates and eigenvalues of the Hamiltonian.

the spin magnetic moment μ_s and a magnetic field of magnitude B oriented along the z-axis is

$$i\hbar \begin{pmatrix} \dot{c}_\uparrow \\ \dot{c}_\downarrow \end{pmatrix} = \begin{pmatrix} -\frac{1}{2}\mu_s\hbar B & 0 \\ 0 & \frac{1}{2}\mu_s\hbar B \end{pmatrix} \begin{pmatrix} c_\uparrow \\ c_\downarrow \end{pmatrix}. \qquad (9.12)$$

The solution is

$$\Psi(t) = \begin{pmatrix} c_\uparrow(0)\exp(i\omega_L t/2) \\ c_\downarrow(0)\exp(-i\omega_L t/2) \end{pmatrix}, \qquad (9.13)$$

where $\omega_L \equiv \mu_s B$ is called the Larmor frequency.

If the state of the system is an eigenstate of the operator $\hat{S}_z$ at $t=0$, it remains so forever and (9.13) is just a particular case of (9.8). However, if at $t=0$ the spin points in the positive x-direction [initial values: $c_\uparrow = c_\downarrow = 1/\sqrt{2}$, see (3.26)], then

$$\Psi(t) = \frac{1}{\sqrt{2}}\begin{pmatrix} 1 \\ 1 \end{pmatrix}\cos\frac{\omega_L t}{2} + i\frac{1}{\sqrt{2}}\begin{pmatrix} 1 \\ -1 \end{pmatrix}\sin\frac{\omega_L t}{2}. \qquad (9.14)$$

The probability of finding the system with spin aligned with the x-axis (or in the opposite direction) is $\cos^2(\omega_L t/2)$ [or $\sin^2(\omega_L t/2)$].

The expectation values of the spin components are

$$\langle\Psi|S_x|\Psi\rangle = \frac{\hbar}{2}\cos(\omega_L t), \quad \langle\Psi|S_y|\Psi\rangle = -\frac{\hbar}{2}\sin(\omega_L t), \quad \langle\Psi|S_z|\Psi\rangle = 0. \qquad (9.15)$$

The spin precesses around the z-axis (the magnetic field axis), with the Larmor frequency ω_L in the negative sense ($x \to -y$). It never aligns itself with the z-axis. Unlike the case in which a definite projection of the angular momentum along the z-axis is well defined (see the discussion of Fig. 5.1 in Sect. 5.1.1), we are describing a true precession here, which is obtained at the expense of the determination of S_z.

If $t \ll 1/\omega_L$, we speak of a transition from the initial state $\varphi_{S_x=\hbar/2}$ to the final state $\varphi_{S_x=-\hbar/2}$ with the probability $\omega_L^2 t^2/4$. In this case, the probability per unit time is linear in time.

9.2.2 Magnetic Resonance

We now add a periodic field along the x- and y-directions, of magnitude B' and frequency ω, to the constant magnetic field of magnitude B pointing along the z-axis. The Hamiltonian reads

$$\hat{H} = -\mu_s B \hat{S}_z - \mu_s B' \left(\cos\omega t \hat{S}_x - \sin\omega t \hat{S}_y\right)$$

$$= -\frac{1}{2}\mu_s\hbar \begin{pmatrix} B & B'\exp(i\omega t) \\ B'\exp(-i\omega t) & -B \end{pmatrix}. \qquad (9.16)$$

Although the solution may be worked out analytically for any value of ω, it turns out that the maximum effect is obtained if this frequency equals the Larmor frequency ω_L. We make this assumption in the derivation below. We also set $\omega' \equiv \mu_s B'$.

We try a solution of the form (9.13), but with time-dependent amplitudes $c_\uparrow(t), c_\downarrow(t)$. The time-dependent Schrödinger equation yields

$$\begin{pmatrix} \dot{c}_\uparrow \\ \dot{c}_\downarrow \end{pmatrix} = \frac{i}{2}\omega' \begin{pmatrix} c_\downarrow \\ c_\uparrow \end{pmatrix} . \tag{9.17}$$

The solution to this equation is

$$\begin{pmatrix} c_\uparrow \\ c_\downarrow \end{pmatrix} = \begin{pmatrix} \cos\dfrac{\omega' t}{2} & i\sin\dfrac{\omega' t}{2} \\ i\sin\dfrac{\omega' t}{2} & -i\cos\dfrac{\omega' t}{2} \end{pmatrix} \begin{pmatrix} c_\uparrow(0) \\ c_\downarrow(0) \end{pmatrix} . \tag{9.18}$$

This result ensures the occurrence of the spin flip: a spin pointing up (down) will eventually point down (up), i.e., it will be flipped. The probabilities that the initial spin is maintained or flipped are (Fig. 9.1)

$$P_{\uparrow\to\uparrow} = P_{\downarrow\to\downarrow} = \cos^2\left(\frac{1}{2}\omega' t\right) ,$$

$$P_{\uparrow\to\downarrow} = P_{\downarrow\to\uparrow} = \sin^2\left(\frac{1}{2}\omega' t\right) . \tag{9.19}$$

For an arbitrary relation between ω and ω_L, the probability of a spin flip is given by

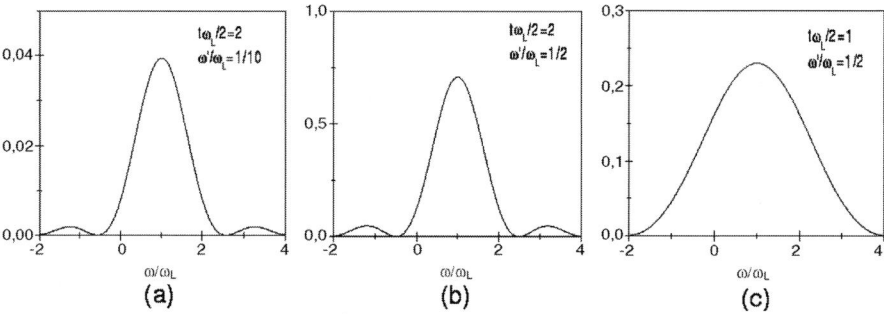

Fig. 9.1. Probability of a spin flip according to (9.20). The resonant behaviour for $\omega = \omega_L$ is apparent in the three figures. Graphs (**a**) and (**b**) differ only in the value of the ratio ω'/ω_L, which is reflected in the value of the probability (*vertical axis*). Graphs (**b**) and (**c**) differ in the value of t. The comparison between the last two graphs anticipates the complementary relation between time and energy [see (9.34) and Sect. 9.5.6]

$$P_{\uparrow\to\downarrow}(t) = \frac{(\omega')^2}{(\omega-\omega_L)^2+(\omega')^2} \sin^2\left[\frac{1}{2}t\sqrt{(\omega-\omega_L)^2+(\omega')^2}\right]. \qquad (9.20)$$

This equation expresses a typical resonance phenomenon (hence the name magnetic resonance): if $\omega \approx \omega_L$, a very weak field B' produces large effects (Fig. 9.1). One cannot treat the interaction with the sinusoidal field as a small perturbation. This would require $|\omega'| \ll |\omega - \omega_L|$, a condition violated in the neighborhood of resonance.

Magnetic resonance is an essential part of processes involving the alignment of spins. It has applications in many branches of physics, such as measuring magnetic moments of particles, including elementary particles, and determining properties of condensed matter. It is also an important tool in quantum computing, medical diagnosis, etc.

9.3 Sudden Change in the Hamiltonian

We consider a time-dependent Hamiltonian $\hat{H}$ such that $\hat{H} = \hat{H}_0$ for $t < 0$ and $\hat{H} = \hat{K}_0$ for $t > 0$, where $\hat{H}_0$ and $\hat{K}_0$ are time-independent Hamiltonians. We know how to solve the problem for these two Hamiltonians:

$$\hat{H}_0 \varphi_i = E_i \varphi_i, \qquad \hat{K}_0 \phi_i = \epsilon_i \phi_i. \qquad (9.21)$$

The system is initially in the state $\varphi_i \exp(-iE_i t/\hbar)$. For $t > 0$, the solution is given by the superposition

$$\Psi = \sum_k c_k \phi_k \exp(-i\epsilon_k t/\hbar), \qquad (9.22)$$

where the amplitudes c_k are time-independent, as is $\hat{K}_0$.

The solution must be continuous in time in order to satisfy a differential equation. Therefore, at $t = 0$,

$$\varphi_i = \sum_k c_k \phi_k \longrightarrow c_k = \langle \phi_k | \varphi_i \rangle. \qquad (9.23)$$

The transition probability is given by

$$P_{\varphi_i \to \phi_k} = |c_k|^2. \qquad (9.24)$$

9.4 Time-Dependent Perturbation Theory

If the Hamiltonian includes both a time-independent term $\hat{H}_0$ and a time-dependent contribution $\hat{V}(t)$, one may still use the expansion (9.11), but with time-dependent amplitudes $[c_i = c_i(t)]$. In that case, the Schrödinger equation for the Hamiltonian $\hat{H}_0 + \hat{V}$ is equivalent to the set of coupled equations

$$i\hbar \sum_i \dot{c}_i \varphi_i \exp(-iE_i t/\hbar) = \hat{V} \sum_i c_i \varphi_i \exp(-iE_i t/\hbar) \, . \tag{9.25}$$

The scalar product with φ_k yields

$$i\hbar \dot{c}_k = \sum_i c_i \langle k|V|i\rangle \exp(i\omega_{ki} t) \, , \tag{9.26}$$

where ω_{ki} is the Bohr frequency and

$$\omega_{ki} = \frac{E_k - E_i}{\hbar} \, . \tag{9.27}$$

This set of coupled equations must be solved together with boundary conditions, such as the value of the amplitudes c_i at $t = 0$. The formulation of the time-dependent problem in terms of the coupled amplitudes $c_i(t)$ is attributed to Dirac.

9.4.1 Transition Amplitudes

The set of coupled equations (9.26) is not easier to solve than (9.5). Therefore, one must resort to a perturbation treatment. As in Sect. 8.1, we multiply the perturbation $\hat{V}(t)$ by the unphysical parameter λ ($0 \le \lambda \le 1$) and expand the amplitudes

$$c_k(t) = c_k^{(0)} + \lambda c_k^{(1)}(t) + \lambda^2 c_k^{(2)}(t) + \cdots \, . \tag{9.28}$$

We impose the initial condition that the system be in the state $\varphi_i^{(0)}(t)$ at $t = 0$. This condition is enforced through the assignment $c_k^{(0)} = \delta_{ki}$, which accounts for terms independent of λ in (9.26).

The perturbation is applied at $t = 0$. Our aim is to calculate the probability of finding the system in another unperturbed eigenstate $\varphi_k^{(0)}$ at time t. The terms linear in λ yield

$$\dot{c}_k^{(1)} = -\frac{i}{\hbar} \langle k|V|i\rangle \exp(i\omega_{ki} t) \, . \tag{9.29}$$

Therefore, the transition amplitudes are given by

$$c_k^{(1)}(t) = -\frac{i}{\hbar} \int_0^t \langle k|V|i\rangle \exp(i\omega_{ki} \tau) d\tau \, . \tag{9.30}$$

The transition probability between the initial state i and the final state k, induced by the Hamiltonian $\hat{V}(t)$, is given in first order of perturbation theory as

$$P_{i \to k}^{(1)}(t) = \left| c_k^{(1)} \right|^2 \, . \tag{9.31}$$

9.4.2 Constant-in-Time Perturbation

Consider matrix elements of the perturbation $\langle k|V|i\rangle$ which do not depend on time in the interval $(0,t)$, and otherwise vanish. The first order amplitude and transition probabilities (9.30) and (9.31) are given by

$$c_k^{(1)} = -\frac{\langle k|V|i\rangle}{\hbar\omega_{ki}} [\exp(i\omega_{ki}t) - 1] , \qquad (9.32)$$

$$P_{i\to k}^{(1)} = \left|\frac{\langle k|V|i\rangle}{\hbar\omega_{ki}}\right|^2 4\sin^2(\omega_{ki}t/2) . \qquad (9.33)$$

This result is common to many first order transition processes. We therefore discuss it in some detail:

- If the final states φ_k belong to a continuous set, the transition probability is proportional to the function $f(\omega) = (4/\omega^2)\sin(\omega t/2)^2$ plotted in Fig. 9.2. The largest peak at $\omega = 0$ has a height proportional to t^2, while the next highest, at $\omega \approx 3\pi/t$, is smaller by a factor of $4/9\pi^2 \approx 1/20$. Therefore, practically all transitions take place for frequencies lying within the central peak, which is characteristic of the phenomenon of resonance. The secondary peaks are associated with diffraction processes.

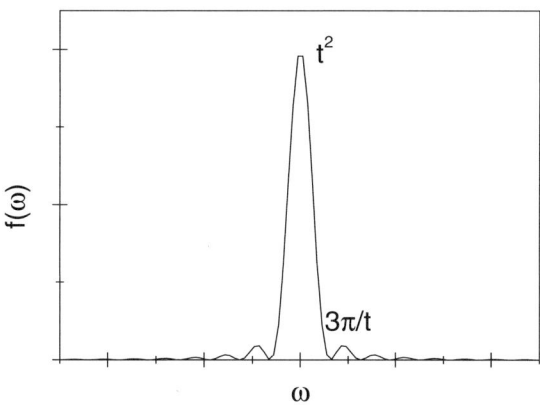

Fig. 9.2. The function $f(\omega)$ as a function of the frequency ω

- The total probability is obtained by integrating over the frequencies. Assuming that the matrix element is not changed within the frequency interval of the main peak, and approximating the surface of the latter by the area of an isosceles triangle of height t^2 and half-base $2\pi/t$, we conclude that the total probability increases linearly with time and that the probability per unit interval of time is constant.

- The energy of an initially excited atomic state is obtained from the frequency of the photon resulting from de-excitation of this state (Sect. 9.5.4). Therefore, the spread shown in Fig. 9.2 changes the notion of the eigenvalue in the case of an unstable state. Instead of a sharp energy, the existence of the spread is associated with an indeterminacy in the energy on the order of

$$\Delta E \geq \hbar \frac{2\pi}{t} \, . \tag{9.34}$$

  This inequality is a manifestation of the uncertainty principle as applied to energy and time variables. There is a similar uncertainty if the energy of the excited states is obtained via a process of absorption of electromagnetic radiation.

- If there is a continuum of final states, we are interested in summing up the probabilities over the set K of these final states ($k \in K$):

$$P^{(1)}_{i \to K} = \int_{E_i - \Delta E/2}^{E_i + \Delta E/2} P^{(1)}_{i \to k} \rho(E_k) \mathrm{d}E_k \, , \tag{9.35}$$

where $\rho(E_k)$ is the density of the final states.[2] Assuming that both $|\langle k|V|i\rangle|^2$ and $\rho(E_k)$ remain constant during the interval ΔE, and that most of the transitions take place within this interval, then

$$P^{(1)}_{i \to K} \approx \frac{4}{\hbar^2} |\langle k|V|i\rangle|^2 \rho(E_k) \int_{-\infty}^{\infty} \mathrm{d}E_k \frac{\sin^2 \omega_{ki} t/2}{\omega_{ki}^2} = \frac{2\pi t}{\hbar} |\langle k|V|i\rangle|^2 \rho(E_k) \, . \tag{9.36}$$

The expression for the transition per unit time is called the Fermi golden rule:

$$\frac{\mathrm{d}P^{(1)}_{i \to K}}{\mathrm{d}t} = \frac{2\pi}{\hbar} |\langle k|V|i\rangle|^2 \rho(E_k) \, . \tag{9.37}$$

9.5 Quantum Electrodynamics for Newcomers

In Chap. 1, we stated that the most important manifestation of the crisis in physics that took place at the beginning of the twentieth century, was the (classical) instability of the motion of an electron circling around the nucleus. For the sake of consistency, we must show that quantum mechanics does indeed solve this problem. However, to do so we must use that beautiful extension of quantum mechanics called quantum electrodynamics. In the following we present a very brief introduction to QED.

We will first consider the electromagnetic field in the absence of charges (light waves). A quadratic expression for energy will be obtained in terms of canonical variables. The theory will be quantized by replacing such variables

---

[2] The density of states is given in (7.21) for the free particle case. A similar procedure is carried out for photons in (9.50).

9.5.1 Classical Description of the Radiation Field

In the absence of charges, the classical electromagnetic vector potential $\boldsymbol{A}(\boldsymbol{r},t)$ satisfies the equation

$$\nabla^2 \boldsymbol{A} = \frac{1}{c^2} \frac{\partial^2 \boldsymbol{A}}{\partial^2 t} \ . \tag{9.38}$$

The vector $\boldsymbol{A}$ may be written as the sum of a transverse and a longitudinal component. The last one can be included within the particle Hamiltonian, since it is responsible for the Coulomb interaction and does not cause the radiation field. The transverse component $\boldsymbol{A}_t(\boldsymbol{r},t)$ satisfies the equation

$$\operatorname{div} \boldsymbol{A}_t = 0 \ . \tag{9.39}$$

It may be expanded in terms of a complete, orthonormal set $\boldsymbol{A}_\lambda(\boldsymbol{r})$ of functions of the coordinates:

$$\boldsymbol{A}_t = \sum_\lambda c_\lambda(t) \boldsymbol{A}_\lambda \ , \tag{9.40}$$

$$\int_{L^3} \boldsymbol{A}_\lambda^* \boldsymbol{A}_{\lambda'} \mathrm{d}v = \delta_{\lambda,\lambda'} \ , \tag{9.41}$$

where we assume a large volume L^3 enclosing the field. Inserting (9.40) in (9.38) and separating variables yields the two equations

$$\frac{\mathrm{d}^2}{\mathrm{d}t^2} c_\lambda + \omega_\lambda^2 c_\lambda = 0 \ , \tag{9.42}$$

$$\nabla^2 \boldsymbol{A}_\lambda + \frac{\omega_\lambda^2}{c^2} \boldsymbol{A}_\lambda = 0 \ , \tag{9.43}$$

where ω_λ is introduced as a separation constant. The solution to the oscillator equation (9.42) is

$$c_\lambda = \alpha_\lambda \exp(-\mathrm{i}\omega_\lambda t) \ , \tag{9.44}$$

with α_λ independent of time. A solution to (9.43) is given by the three-dimensional generalization (7.17) of (4.41). Periodic boundary conditions are assumed, so that

$$\boldsymbol{A}_\lambda = \frac{1}{L^{3/2}} \boldsymbol{v}_\lambda \exp(\mathrm{i}\boldsymbol{k}_\lambda \cdot \boldsymbol{r}) \ , \qquad k_{\lambda i} = 2\pi n_{\lambda i}/L \ . \tag{9.45}$$

There are two independent directions of polarization $\boldsymbol{v}_\lambda$, since (9.39) implies $\boldsymbol{v}_\lambda \cdot \boldsymbol{k}_\lambda = 0$.

We construct the electric field

$$\boldsymbol{E} = -\frac{\partial}{\partial t}\boldsymbol{A}_t = -\sum_\lambda \omega_\lambda c_\lambda \boldsymbol{A}_\lambda \ . \qquad (9.46)$$

The total field energy is expressed as

$$\begin{aligned} U &= \frac{1}{2}\int_V \left(\epsilon_0|\boldsymbol{E}|^2 + \mu_0|\boldsymbol{B}|^2\right)\mathrm{d}V = \int_V \epsilon_0|\boldsymbol{E}|^2\mathrm{d}V \\ &= \epsilon_0 \sum_\lambda \omega_\lambda^2 c_\lambda^* c_\lambda \\ &= \sum_\lambda \hbar\omega_\lambda a_\lambda^* a_\lambda \ , \end{aligned} \qquad (9.47)$$

where the substitution

$$c_\lambda = c\sqrt{\frac{\hbar}{\epsilon_0 \omega_\lambda}} a_\lambda$$

has been made. Note that since the vector field has dimension $\mathrm{k\,m\,s^{-1}\,C^{-1}}$, the amplitudes c_λ have dimension $\mathrm{k\,m^{5/2}\,s^{-1}\,C^{-1}}$ and the amplitudes a_λ have dimension 1.

9.5.2 Quantization of the Radiation Field

We have obtained an expression for the energy of the radiation field that is quadratic in the amplitudes a_λ^*, a_λ. Quantization is achieved by replacing these amplitudes by the creation and annihilation operators a_λ^+, a_λ, satisfying the commutation relation (3.37) [equivalent to (2.8)]. We thus obtain the Hamiltonian[3]

$$\hat{H} = \sum_\lambda \hbar\omega_\lambda\, a_\lambda^+ a_\lambda \ . \qquad (9.48)$$

This Hamiltonian implies that:

- The radiation field is made up of an infinite number of harmonic oscillators. The state of the radiation field is described by all the occupation numbers n_λ.
- In agreement with Planck's postulate, each oscillator has an energy which is a multiple of $\hbar\omega_\lambda$. The energy of the field is the sum of the energies of each oscillator.
- Since the radiation field is a function defined at all points of space and time, the number of canonical variables needed for its description is necessarily infinite. However, by enclosing the field within the volume L^3, we have succeeded in transforming this infinity into a denumerable infinity.

---

[3] We ignore the ground state energy of the radiation field.

- In the absence of any interaction between particles and radiation field, vector states may be written as products of the two Hilbert subspaces, and the energy $E_{b,n_1,n_2,...}$ is the sum of particle and radiation terms (7.1):

$$\Psi_{b,n_1,n_2,...} = \varphi_b(\text{particles}) \times \Pi_\lambda \frac{1}{\sqrt{n_\lambda!}} (a_\lambda^+)^{n_\lambda} \varphi_0 ,$$

$$E_{b,n_1,n_2,...} = E_b + \sum_\lambda \hbar\omega_\lambda n_\lambda . \tag{9.49}$$

- The number of states up to a certain energy $n(E_\lambda)$ and per unit interval of energy $\rho(E_\lambda)$ for each independent direction of polarization are[4]

$$n(E_\lambda) = \frac{L^3 k_\lambda^3}{6\pi^2} = \frac{L^3 E_\lambda^3}{6\pi^2 \hbar^3 c^3} , \quad \rho(E_\lambda) = \frac{\partial n}{\partial E_\lambda} = \frac{L^3 \omega_\lambda^2}{2\pi^2 \hbar c^3} . \tag{9.50}$$

- The Hermitian, quantized, vector potential reads

$$\hat{A}_t = \frac{1}{2} \sum_\lambda \sqrt{\frac{\hbar}{\epsilon_0 L^3 \omega_\lambda}} \left[a_\lambda \boldsymbol{v}_\lambda \exp(i\boldsymbol{k}_\lambda \cdot \boldsymbol{r}) + a_\lambda^+ \boldsymbol{v}_\lambda \exp(-i\boldsymbol{k}_\lambda \cdot \boldsymbol{r}) \right] . \tag{9.51}$$

9.5.3 Interaction of Light with Particles

In the presence of an electromagnetic field, the momentum $\hat{\boldsymbol{p}}$ of the particles[5] is replaced in the Hamiltonian by the effective momentum [50]

$$\hat{\boldsymbol{p}} \longrightarrow \hat{\boldsymbol{p}} - e\hat{\boldsymbol{A}}_t , \tag{9.52}$$

$$\frac{1}{2M}\hat{\boldsymbol{p}}^2 \longrightarrow \frac{1}{2M}\hat{\boldsymbol{p}}^2 + \hat{V} + \cdots , \quad \hat{V} = \sqrt{\frac{\alpha 4\pi\epsilon_0 \hbar c}{M^2}} \hat{\boldsymbol{A}}_t \cdot \hat{\boldsymbol{p}} . \tag{9.53}$$

The various ensuing processes may be classified according to the associated power of the fine structure constant α. The smallness of α (Table 13.1) ensures the convergence of perturbation theory. The linear, lowest order processes require only the perturbation term $\hat{V}$. This causes transitions in the unperturbed system, particle + radiation, by changing the state of the particle and simultaneously increasing or decreasing the number of field quanta by one unit (emission or absorption processes, respectively).

We apply the perturbation theory developed in Sect. 9.4.2. Since the radiation field has a continuous spectrum, a transition probability per unit

---

[4] These expressions have been derived using a similar procedure to the one used to obtain (7.21). A factor of 2, which was included in (7.21) due to spin, is not needed in (9.50). It reappears in (9.56) below, where the two directions of polarization are taken into account.

[5] $[\hat{\boldsymbol{p}}, \hat{\boldsymbol{A}}_t] = 0$ because of (9.39).

time (9.37) is obtained. Energy is conserved for the total system within the time–energy uncertainty relation.

According to (9.51) and (9.53), the matrix elements of the perturbation read

$$\langle b(n_\lambda + 1)|V|an_\lambda\rangle = K_\lambda \sqrt{n_\lambda + 1},$$
$$\langle b(n_\lambda - 1)|V|an_\lambda\rangle = K_\lambda \sqrt{n_\lambda}, \quad (9.54)$$

where K_λ is given by

$$\begin{aligned} K_\lambda &= \frac{\hbar}{M}\sqrt{\frac{\alpha\pi c}{L^3\omega_\lambda}}\langle b|(\boldsymbol{v}_\lambda\!\cdot\!\boldsymbol{p})\exp(\pm i\boldsymbol{k}_\lambda\!\cdot\!\boldsymbol{r})|a\rangle \\ &\approx \frac{\hbar}{M}\sqrt{\frac{\alpha\pi c}{L^3\omega_\lambda}}\langle b|\boldsymbol{v}_\lambda\!\cdot\!\boldsymbol{p}|a\rangle \\ &= i\hbar\omega_\lambda\sqrt{\frac{\alpha\pi c}{L^3\omega_\lambda}}\langle b|\boldsymbol{v}_\lambda\!\cdot\!\boldsymbol{r}|a\rangle. \end{aligned} \quad (9.55)$$

We have neglected the exponential within the matrix element, on the basis of the estimate: $\langle k_\lambda r\rangle \approx \omega_\lambda a_0/c = O(10^{-4})$, for $\hbar\omega_\lambda \approx 1$ eV. The third line is derived using the relation $\hat{p} = (iM/\hbar)[\hat{H},\hat{x}]$ (Problem 9 of Chap. 2).

We next work out the product appearing in the golden rule (9.37):

$$\frac{2\pi}{\hbar}|K_\lambda|^2\rho(E_\lambda) = \frac{\alpha|\omega_\lambda|^3}{c^2}|\langle b|\boldsymbol{v}_\lambda\!\cdot\!\boldsymbol{r}|a\rangle|^2$$
$$\longrightarrow \frac{2\alpha|\omega_\lambda|^3}{3c^2}|\langle b|\boldsymbol{r}|a\rangle|^2, \quad (9.56)$$

where we have summed over the two final polarization directions and averaged over them.

9.5.4 Emission and Absorption of Radiation

The transition probability per unit time is given by

$$\begin{aligned} \frac{dP^{(1)}_{an_\lambda\to b(n_\lambda-1)}}{dt} &= \frac{2\alpha|\omega_\lambda|^3}{3c^2}|\langle b|\boldsymbol{r}|a\rangle|^2\, \bar{n}_\lambda, \\ \frac{dP^{(1)}_{an_\lambda\to b(n_\lambda+1)}}{dt} &= \frac{2\alpha|\omega_\lambda|^3}{3c^2}|\langle b|\boldsymbol{r}|a\rangle|^2\,(\bar{n}_\lambda + 1), \end{aligned} \quad (9.57)$$

for absorption and emission processes, respectively. Here $\bar{n}_\lambda$ is the average number of photons of a given frequency.

The probability of absorbing a photon is proportional to the intensity of the radiation field present before the transition. This intensity is represented by $\bar{n}_\lambda$. This is to be expected. However, the probability of emission consists of two terms: the first one also depends on the intensity of the radiation field

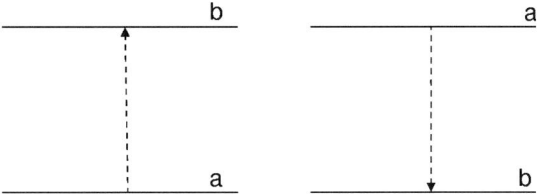

Fig. 9.3. Absorption (left) and emission (right) of electromagnetic radiation

(induced emission); the second term, independent of the field intensity, allows the atom to decay from an excited state in vacuo (spontaneous emission).

The ratio $(\bar{n}_\lambda+1)/\bar{n}_\lambda$ is needed to preserve the correct thermal equilibrium of the radiation with a gas: in a gas at temperature T, the number of atoms in the states a, b is given by $\exp(-E_a/k_\mathrm{B}T)$ and $\exp(-E_b/k_\mathrm{B}T)$, respectively. The condition for equilibrium is

$$P_\mathrm{emission} \exp(-E_a/k_\mathrm{B}T) = P_\mathrm{absorption} \exp(-E_a/k_\mathrm{B}T) , \tag{9.58}$$

which yields

$$\bar{n}_\lambda = 1/\left[\exp(\hbar\omega_{ab}/k_\mathrm{B}T) - 1\right] . \tag{9.59}$$

From this deduction of Planck's law, Einstein showed the need for spontaneous and induced emission in quantum theory [55].

9.5.5 Selection Rules

We now focus our attention on the particle matrix elements. The transition probabilities are also proportional to the squared modulus of the matrix elements $|\langle b|\boldsymbol{r}|a\rangle|^2$. Therefore, transition rates give information about the value of non-diagonal matrix elements. (Since Chap. 2, we know that diagonal matrix elements represent averages obtained in measurements of the eigenvalues of physical observables.)

Let l_b, π_b (l_a, π_a) be the orbital angular momentum and parity quantum numbers of the final (initial) state. Conservation of angular momentum requires the orbital angular momentum of the final state to equal the vector sum of the initial angular momentum and that of the radiation. The latter manifests itself through the operator $\hat{\boldsymbol{r}}$ in (9.57), which can be expressed as a sum of terms proportional to the spherical harmonics Y_{1m_l}. Therefore, the matrix elements must satisfy the selection rule $l_a + 1 \geq l_b \geq |l_a - 1|$ (5.33), or $\Delta l = 0, \pm 1$.

Since the operator $\boldsymbol{r}$ is odd under the parity operation (5.13), the non-vanishing of this matrix element also requires the initial and final states to carry a different parity, $\pi_a \pi_b = -1$. The combination of the conservation rules associated with orbital angular momentum and parity are condensed in the selection rule

$$\Delta l = \pm 1 \,, \tag{9.60}$$

which defines allowed transitions. Forbidden (i.e., non-allowed) transitions may also occur, but are much weaker than the allowed ones. Their relative intensity may be estimated on the base of the expectation value of the neglected terms in (9.55).

9.5.6 Mean Lifetime and Energy–Time Uncertainty Relation

A system can make a transition to another state only if it is not in an eigenstate of the Hamiltonian. This is the case of the examples studied in Sects. 9.2 and 9.4, where the transition probability dP/dt per unit time was calculated for a single system. If there are $\mathcal{N}$ similar systems present (for instance, $\mathcal{N}$ atoms), one cannot ascertain when a particular system will decay. The total rate of change is given by

$$\frac{d\mathcal{N}}{dt} = -\mathcal{N}\frac{dP}{dt} \,. \tag{9.61}$$

Therefore,

$$\mathcal{N} = \mathcal{N}_0 \exp\left(-\frac{dP}{dt}t\right) = \mathcal{N}_0 \exp(-t/\tau) \,, \qquad \tau = \left(\frac{dP}{dt}\right)^{-1} \,. \tag{9.62}$$

The constant τ is called the mean lifetime. It is the time required for the reduction of the size of the population by a factor of $1/e$.

As in the case of the uncertainty relation concerning position and momentum (Sect. 2.6), the uncertainty relation (9.34) has to be applied to measurements of many identical systems. The replacement of the time t in (9.34) by the mean lifetime τ relates the minimum spread of energy to lifetimes. A short mean lifetime implies a broad peak, and vice versa.

The mean lifetime of the excited state φ_{210} in the hydrogen atom ($\hbar\omega = 10.2$ eV) may be obtained[6] from (9.57) and (9.62). The result yields $\tau = 0.34 \times 10^{-9}$ s. Does this represent a short or a long time? In fact it is a long time, since this mean lifetime has to be compared with the period of the emitted radiation $\mathcal{T} = 2\pi/\omega = 0.41 \times 10^{-15}$ s. The mean lifetime is associated with a spread in energy of 1.23×10^{-5} eV, which is much smaller than the excitation energy. We can see now how effectively the great crisis of early twentieth century physics was resolved.

---

[6] An order of magnitude of τ may be obtained by equating the rate of radiation of an oscillating classical dipole with the ratio between the emitted energy $\hbar\omega$ and the mean lifetime

$$\frac{\omega^4 D_0}{3c^2} = \frac{\hbar\omega}{\tau} \longrightarrow \tau = \frac{3\hbar c^2}{\omega^3 D_0} \,.$$

The amplitude of the dipole oscillation is approximated by $D_0 \approx -ea_0$. If the transition energy is assumed to be 10 eV, we obtain an estimated value $\tau = O(10^{-10}$ s$)$.

Problems

Problem 1. At $t = 0$, a state is given by the linear combination of the two lowest states of a linear, infinite square well potential of width a:

$$\Psi(t=0) = \frac{1}{\sqrt{3}}\varphi_1 - i\sqrt{\frac{2}{3}}\varphi_2 \, .$$

1. Write the wave function at time t.
2. Calculate the probability of finding the particle in the second half of the well.

Problem 2. Use the time-dependent Schrödinger equation to show that Newton's second law is obeyed on average in quantum mechanics (Ehrenfest theorem). Hint: calculate

$$\frac{\mathrm{d}\langle\Psi(t)|p|\Psi(t)\rangle}{\mathrm{d}t},$$

as in (4.14).

Problem 3. In the state (9.14), calculate the amplitude of the eigenstate of the operator $\hat{S}_y$, with spin pointing in the positive direction.

Problem 4. A particle is in the ground state of an infinite linear square well potential. What is the probability of finding it in the $n = 1, 2, 3$ states when the wall separation is suddenly doubled by displacing the right wall?

Problem 5. Calculate the probability of a spin flip in the first order of perturbation theory. Assume the Hamiltonian (9.16) and $|\omega'/(\omega - \omega_\mathrm{L})| \ll 1$. Compare with the exact result (9.20).

Problem 6. A particle in the ground state of a linear harmonic oscillator interacts with a projectile through an interaction of the form $V_0\delta(u - vt/x_\mathrm{c})$.

1. Express the amplitude for the transition to the first excited state as an integral over the time interval $t_1 \leq t \leq t_2$.
2. Calculate the probability of this transition for $t_1 = -\infty$, $t_2 = \infty$.

Problem 7. The Hamiltonian

$$\hat{V}(t) = \frac{V_0}{\hbar^2}\hat{\boldsymbol{S}}_1 \cdot \hat{\boldsymbol{S}}_2 \cos(\omega t)$$

acts on a two-spin system with a total projection of $m_s = 0$.

1. Find the time-dependent solution.

2. Do the same with $\hat{V}(t)$ acting on the Bell state Ψ_{B_0} (10.25). Hint: Compare (7.12) and Problem 9 of Chap. 6 and try

$$\Psi(t) = \cos\theta \exp(i\phi)\chi_0^1 + \sin\theta \exp(-i3\phi)\chi_0^0 .$$

Problem 8. What is the probability of exciting a linear harmonic oscillator from the ground state to the first excited state, assuming that a perturbation $V = Kx$, acting for an interval of t, is added to the oscillator Hamiltonian?

Problem 9.

1. Obtain the expression for the second order amplitudes $c_k^{(2)}(t)$ if the perturbation is constant in time.
2. Calculate the probability of a transition to the second excited state for the same problem as in Problem 8.

Problem 10.

1. Assuming that 2 keV is the spread in energy of a nuclear state, calculate its mean lifetime and compare it with the time it takes for a beam of light to cross the nucleus (nuclear size $\approx 10^{-12}$ m).
2. Do the same for a meson with a spread of 200 MeV and a beam of light crossing a proton (proton size $\approx 10^{-17}$ m).

Problem 11. Calculate the ratio between the populations of the states φ_{210} and φ_{310} if hydrogen atoms in their ground state are illuminated with white light.

Problem 12.

1. Calculate the ratio between the intensities of photons de-exciting the state φ_{310} of the hydrogen atom.
2. Calculate the mean lifetime of this state.
3. Calculate the width of this state.

10 Entanglement and Some Recent Applications of Quantum Mechanics

It's not your grandfather's quantum mechanics. [62]

There are unique features of quantum mechanics that only appear for two or more identical particles. Among them, the concept of entanglement is not only fundamental in the discussion of the validity of quantum mechanics (Sects. 12.4.2 and 12.4.3), but it has also become, during the last twenty years, a central tool in the emerging field of quantum information. We will not attempt to give a complete description of these recent developments. Rather, we will use the respectable knowledge of quantum mechanics which readers should now have in order to illustrate these new uses with pertinent examples.

All these applications make use of two-dimensional Hilbert spaces. Therefore the considerations on these spaces given in Sect. 5.2.2 will be assumed here.

10.1 Entanglement

If the state representing two or more quantum systems cannot be expressed as a product of separate states for each system, the state is said to be entangled. Examples of two-particle entangled states appear in (7.5) and (7.6).

Let us perform some thought experiments: two particles 1 and 2 are emitted simultaneously, from the same source, in different directions. The source includes a filter aligned with the laboratory z-axis. Each particle may be detected by an observer provided with another filter. The same orientation β, relative to the laboratory frame, holds for both observers' filters. The down channel is also blocked in both filters. Thus $\cos^2(\beta/2)$ represents the probability that a particle goes through a given filter, and $\sin^2(\beta/2)$ the probability that it gets absorbed.

If two particles are emitted in the entangled state

$$\frac{1}{\sqrt{2}}\left[\varphi_\uparrow(1)\varphi_\uparrow(2) + \varphi_\downarrow(1)\varphi_\downarrow(2)\right],$$

then:

- A measurement of particle 1 destroys the entangled state. Particle 2 assumes the same state as the one into which particle 1 was projected by the measurement.

- The correlation is 100%, regardless of the filter orientation β.
- This result is also independent of the initial direction of the z-axis. Replacing $\varphi_\uparrow, \varphi_\downarrow$ by linear orthonormal combinations $\phi_\uparrow, \phi_\downarrow$ generated by rotations around the y-axis, yields the entangled state

$$\frac{1}{\sqrt{2}}\left[\phi_\uparrow(1)\phi_\uparrow(2) + \phi_\downarrow(1)\phi_\downarrow(2)\right].$$

A measurement of particle 1 would project particle 2 into the same state as particle 1.
- The correlation takes place regardless of the distance between the two particles: particle 2 (non-locally) acquires information about the result of the measurement of particle 1.
- The comparison with the (non-entangled) product state $\varphi_\uparrow(1)\varphi_\uparrow(2)$ is illuminating: here the probability that the two particles get through is $\cos^4(\beta/2)$, while the probability that both are absorbed is $\sin^4(\beta/2)$. Therefore the probability that both observers find the same result is $1 - (1/2)\sin^2\beta$. If $\beta = \pi/2$, this last probability has the value $1/2$, the same classical value as for two independent players tossing coins.

Entanglement constitutes a profound, non-classical correlation between two (or more) quantum systems. The members of entangled systems do not have their own individual quantum states. Only the total system is in a well defined state. This is fundamentally unlike anything in classical physics.

The notion of entanglement will appear in the discussion of the EPR paradox (Sect. 12.4.2), in the experimental tests of the validity of quantum mechanics (Sect. 12.4.3), in the interpretation of measurements in terms of decoherence (Sect. 11.2), in the description of teleportation (Sect. 10.3), and in the presentation of quantum computation (Sect. 10.4).

A complete set of entangled, two-particle states is given in Sect. 10.5*.

10.2 Quantum Cryptography

Traditional strategies for keeping secrets in the distribution of cryptographic keys depend on human factors, so their safety is difficult to assess. As a consequence, they have been replaced to a large extent by cryptosystems. A cryptographic key is transmitted through a random series of the numbers 0 and 1. Their present safety is due to the fact that no fast algorithms can work out the decomposition of a large number in prime factors. However this statement, which is valid for classical computation, may no longer be true with the advent of quantum computation (see Sect. 10.4). Hence the continuing interest in exploring safer systems for transmission of cryptographic keys. In this section, we show that the quantum key distributions are impossible to break, and that this impossibility arises from fundamental quantum laws.

10.2 Quantum Cryptography

The best known protocol is called BB84 [63]: every actor involved is provided with a filter of the type discussed in Sect. 2.4. The encoder (usually named Alice) can send particles that are in an eigenstate of either $\hat{S}_z$ or $\hat{S}_x$. We label the corresponding states by $\uparrow z, \downarrow z, \uparrow x, \downarrow x$. The decoder (frequently called Bob) may orient his detection equipment along either the z- or the x-direction. For instance, if Alice has sent 3 qubits that are polarized according to $\uparrow z, \uparrow x$ and $\downarrow x$, and if Bob aligns his apparatus in the z-direction for the first qubit and the x-direction for the last two, he will detect intensities 1, 1, and 0 with certainty. These are the good qubits, i.e., those sent and measured with both pieces of apparatus along the same orientation. If Alice sends a qubit in the $\downarrow z$ state while Bob orients his equipment towards the x-direction, he may detect the intensities 1 or 0 with equal probability. Bad qubits are those in which the sender and receiver apparatus have different orientations.

For each qubit sent, Alice records the eigenvalue as well as the orientation of her filter. Bob selects his orientations at random and informs Alice of them. With this knowledge, Alice can tell Bob which ones are good qubits, the ones which are kept in order to codify the message. Both messages from Bob and Alice can even be made over an open phone, since they carry no information useful for a third party.

Let us assume that there is an eavesdropper, usually called Eve. Eve cannot be prevented from eavesdropping, but if she does, Alice and Bob will know: let us assume that Alice has sent a qubit in a $\downarrow z$ state, that Eve's apparatus is in orientation x and Bob's in orientation z. The action of Eve increases the probability that Bob will detect the particle from 0 to 1/2 $[= |\langle \downarrow z| \uparrow x \rangle \langle \uparrow x | \uparrow z \rangle|^2$, see (3.26)]. Bob chooses a random subset of the good qubits that he has retained, and communicates them to Alice, also publicly. Alice may find discrepancies between her notes and Bob's message. If she does not, all good qubits constitute a perfect secret between Alice and Bob.

Quantum cryptography applies the rule that quantum states are perturbed by the act of measurement, unless the observer knows in advance what observables can be measured without being perturbed (Sect. 2.5). Eve cannot succeed without knowing the base common to both Alice and Bob.

Experiments of this kind have been made with polarized photons, using optical fibers and also in free space. The experimental problems are mainly associated with the weakness of light pulses, since the individual photons must be separated in time. Experiments over 50 km have been performed with optical fibers and over 1.6 km in open air [64].

Although it has not been employed in this section, entanglement has been used to create pairs of identical cryptographic keys.

10.3 Quantum Teleportation

Alice and Bob are at a macroscopic distance from each other. Alice's particle is initially in the Ψ_c state

$$\Psi_c = c_\uparrow \varphi_\uparrow + c_\downarrow \varphi_\downarrow \equiv \begin{pmatrix} c_\downarrow \\ c_\uparrow \end{pmatrix}, \qquad |c_\uparrow|^2 + |c_\downarrow|^2 = 1. \tag{10.1}$$

The objective is to put Bob's particle in the same state, but without transporting the particle or sending any classical information about it.

Alice and Bob start by each taking one of the two qubits which have been prepared, for instance in the Bell state φ_{B_0} (10.25). Alice now has two qubits, one in the state Ψ_c and the other in the Bell state. The three-qubit state can be written as

$$\Psi = \Psi_c \varphi_{B_0} = \frac{1}{\sqrt{2}} \left(c_\uparrow \varphi_{\uparrow\uparrow\uparrow} + c_\uparrow \varphi_{\uparrow\downarrow\downarrow} + c_\downarrow \varphi_{\downarrow\uparrow\uparrow} + c_\downarrow \varphi_{\downarrow\downarrow\downarrow} \right) \tag{10.2}$$

$$= \frac{1}{\sqrt{2}} \left(c_\uparrow \varphi_{\uparrow\uparrow} \varphi_\uparrow + c_\uparrow \varphi_{\uparrow\downarrow} \varphi_\downarrow + c_\downarrow \varphi_{\downarrow\uparrow} \varphi_\uparrow + c_\downarrow \varphi_{\downarrow\downarrow} \varphi_\downarrow \right), \tag{10.3}$$

where the qubit taken by Bob from the Bell state has been explicitly separated in the last line. Inverting (10.25) and replacing the two-qubit states by Bell states yields

$$\Psi = \frac{1}{2} \left[\varphi_{B_0} \begin{pmatrix} c_\downarrow \\ c_\uparrow \end{pmatrix} + \varphi_{B_1} \begin{pmatrix} c_\downarrow \\ -c_\uparrow \end{pmatrix} + \varphi_{B_2} \begin{pmatrix} c_\uparrow \\ c_\downarrow \end{pmatrix} + \varphi_{B_3} \begin{pmatrix} -c_\uparrow \\ c_\downarrow \end{pmatrix} \right]. \tag{10.4}$$

Alice now filters her two qubits into a well defined Bell state (see Sect. 10.5*). Simultaneously, Bob's qubit is also projected into a well defined state, but Bob ignores the relation between this state and the initial state Ψ_c. Bob needs to know in which Bell state the system has collapsed in order to reconstruct the original qubit. This information must be provided by Alice by conventional means, i.e., at a speed less than or equal to the velocity of light.

Suppose for instance that, instead of going through the previous procedure, Alice constructs the state Ψ_c by filtering the spin, and sends the information about the alignment axis to Bob, who can thus filter the particle in the same direction. Are there still advantages in teleportation? The answer is affirmative, for the following reasons:

- The teleported state Ψ_c might not be known by Alice. If she attempts to measure it, the state of the qubit could be changed.
- Bob receives complete information about Alice's qubit at the expense of that qubit: in quantum teleportation the original qubit is destroyed. This is a manifestation of the no-cloning theorem (Sect. 10.6*).
- The qubit Ψ_c is determined by the amplitudes $c_\uparrow, c_\downarrow$, for which the transmission time increases with the required precision. Now the results of the quantum experiments are discrete numbers. In quantum teleportation, discrete information about the Bell state is transformed into continuous information about the state of the qubit.

Quantum teleportation was discovered in 1993 [66]. It was observed for the first time in 1997 with entangled photons [67].

10.4 Quantum Computation

In this section we present:

- a brief discussion on the conceptual framework of quantum computation [25],
- the Hamiltonian acting upon a system of qubits,
- the most typical quantum gates (unitary transformations) operating on these qubits and their realization in terms of the Hamiltonian (Sect. 10.7*),
- the factorization of the number 15, as a 'best typical' example.

10.4.1 Conceptual Framework

Assume for a moment that the Hilbert space is restricted to the pure basis states. For a single qubit, the only available states would thus be the bits $\varphi_\uparrow$ and $\varphi_\downarrow$. For n qubits, there are 2^n orthogonal vectors φ_i in a space of 2^n dimensions. Classical computation operates in this space. Linear combinations of vectors are not allowed. Therefore, the only operations that can be performed are permutations between the basis states (unless the size of the space is changed).

Quantum mechanics allows for superpositions Ψ of the basis vectors φ_i with complex amplitudes c_i (2.3). Quantum operations are only limited by the requirement that they be unitary, i.e., that they preserve the norm of the state. Therefore, classical states and classical operations constitute sets of vanishingly small size relative to those sets encompassing quantum states and quantum operations. Thus, quantum computation offers a wealth of new possibilities: all kinds of interference effects may take place in the much larger space, much faster calculations can be performed, if all components of the state vector work in parallel, and so on.

However, this promising picture is limited by the fact that it is very difficult to extract anything from the state vector Ψ, in spite of the immense amount of information that it carries. In fact, the only way is to perform a measurement, which relates Ψ to a single probability $|c_i|^2$. Moreover, it is not possible to obtain additional information about Ψ from a second measurement, since the system collapses into the state φ_i after the first one. Therefore, one should produce transformations that lead to a state Ψ', in which very few amplitudes c'_i do not vanish.

Consequently, a quantum computation starts with the preparation of the system in some initial state φ_0 (i.e, with a measurement) and ends with another measurement in the final state φ_f (see Sect. 2.4). Between these two operations, a quantum computer, by taking advantage of interference and entanglement, can perform in a reasonable time some tasks that would take a prohibitively long time using classical computers. In particular, the factorization of a number into its prime components is transformed from a problem in which time increases exponentially, to a problem in which it increases only polynomially (Sect. 10.4.3).

10.4.2 Quantum Gates

A quantum gate is a device that performs a unitary transformation on selected qubits at a certain time. A quantum network is a device consisting of quantum gates that are synchronized in time.

From the engineering point of view, it is practical to restrict the transformations to those operations that may be expressed as the product of operations on one- and two-qubit systems. Manipulations of a single qubit may be performed by controlling a magnetic field at its site (Sect. 9.2). Simultaneous manipulations of two qubits require an interaction between them. Therefore we use a controlling Hamiltonian

$$\hat{H}_{\text{ctr}} = -\mu_s \sum_i^N \boldsymbol{B}^{(i)}(t) \cdot \hat{\boldsymbol{S}}^{(i)} + \sum_{\substack{a,b \\ i \neq j}} J_{ab}^{(i,j)}(t) \hat{S}_a^{(i)} \hat{S}_b^{(j)} \,, \qquad (10.5)$$

where summation over space indices $a, b = x, y, z$ is understood (see Problem 9 in Chap. 6 and Problem 7 in Chap. 9). This Hamiltonian satisfies the requirements for controlling a quantum computer. In fact, it even exceeds them. Interaction with the measurement device and with the environment should also be taken into account.

One-Qubit Systems

We use the familiar column vector notation

$$\varphi_0 \equiv \begin{pmatrix} 1 \\ 0 \end{pmatrix}, \qquad \varphi_1 \equiv \begin{pmatrix} 0 \\ 1 \end{pmatrix}. \qquad (10.6)$$

A qubit is manipulated by acting with the first term in (10.5). Switching on the x- or z-component of the magnetic field during a time τ introduces the transformations[1] (5.10):

$$\mathcal{U}_x(\alpha) = \exp(\mathrm{i}\mu_s B_x \tau \hat{S}_x) = \begin{pmatrix} \cos(\alpha/2) & \mathrm{i}\sin(\alpha/2) \\ \mathrm{i}\sin(\alpha/2) & \cos(\alpha/2) \end{pmatrix}, \qquad (10.7)$$

$$\mathcal{U}_z(\beta) = \exp(\mathrm{i}\mu_s B_z \tau \hat{S}_z) = \begin{pmatrix} \exp(\mathrm{i}\beta/2) & 0 \\ 0 & \exp(-\mathrm{i}\beta/2) \end{pmatrix}, \qquad (10.8)$$

---
[1] We have used the fact that $\hat{S}_\nu^2 = \hbar^2 \mathcal{I}/4$ (5.22) in the expansion of the exponent.

where $\alpha = \mu_s B_x \tau/\hbar$, $\beta = \mu_s B_z \tau/\hbar$.

Particularly useful gates are the Hadamard gate $\mathcal{U}_\mathrm{H}$ and the phase gate $\mathcal{U}_\phi(\beta)$, which transform the one-qubit states through the operations

$$\mathcal{U}_\mathrm{H}\varphi_J = \frac{1}{\sqrt{2}} \sum_{K=0}^{K=1} \exp(\mathrm{i}JK\pi)\varphi_K \,, \qquad \mathcal{U}_\phi(\beta)\varphi_J = \exp(\mathrm{i}J\beta)\varphi_J \,, \qquad (10.9)$$

where $J = 0, 1$ (see Sect. 10.7*).[2] The Hadamard gate transforms the states (10.6)

$$\mathcal{U}_\mathrm{H}\varphi_0 = \frac{1}{\sqrt{2}}(\varphi_0 + \varphi_1) \,, \qquad \mathcal{U}_\mathrm{H}\varphi_1 = \frac{1}{\sqrt{2}}(\varphi_0 - \varphi_1) \,, \qquad (10.10)$$

and the phase gate adds a phase to the state φ_1. These two operations are sufficient to construct any unitary operation on a single qubit, since

$$\mathcal{U}_\phi(\eta + \pi/2)\mathcal{U}_\mathrm{H}\mathcal{U}_\phi(\beta)\mathcal{U}_\mathrm{H}\varphi_0 = \varphi_0 \cos\frac{\beta}{2} + \varphi_1 \exp(\mathrm{i}\eta)\sin\frac{\beta}{2} \,, \qquad (10.11)$$

which is the most general form for a qubit.

The realization of the Hadamard and phase gates by means of the field term in the controlling Hamiltonian (10.5) is given in Sect. 10.7*.

Two-Qubit Systems

The two-qubit states $\varphi_J^{(2)}$, with $J = 0, 1, 2, 3$, can be represented as products of single qubits or in column vector notation

$$\varphi_0^{(2)} = \varphi_0\varphi_0 \equiv \begin{pmatrix} 1 \\ 0 \\ 0 \\ 0 \end{pmatrix} \,, \qquad \varphi_1^{(2)} = \varphi_0\varphi_1 \equiv \begin{pmatrix} 0 \\ 1 \\ 0 \\ 0 \end{pmatrix} \,,$$

$$\varphi_2^{(2)} = \varphi_1\varphi_0 \equiv \begin{pmatrix} 0 \\ 0 \\ 1 \\ 0 \end{pmatrix} \,, \qquad \varphi_3^{(2)} = \varphi_1\varphi_1 \equiv \begin{pmatrix} 0 \\ 0 \\ 0 \\ 1 \end{pmatrix} \,. \qquad (10.12)$$

It is customary to think of the first qubit in the product representation as the control qubit, and the second as the target qubit.

Useful gates acting on two-qubit systems are the controlled-NOT gate $\mathcal{U}_\mathrm{CNOT}$ and the controlled-phase gate $\mathcal{U}_\mathrm{CB}(\phi)$. These two gates apply a single-qubit transformation to the target qubit if the control qubit is in the state φ_1, and do nothing if the control qubit is in the state φ_0. Thus

---

[2] We keep the quantum mechanical notation previously used in this text. In computation texts, the Hadamard gate is denoted by H, successive transformations are read from left to right, and so on.

$$\mathcal{U}_{\mathrm{CNOT}}\varphi_J\varphi_K = \varphi_J\varphi_{J\oplus K}\,, \qquad \mathcal{U}_{\mathrm{CB}}(\phi)\varphi_J\varphi_K = \exp(\mathrm{i}JK\phi)\varphi_J\varphi_K\,, \qquad (10.13)$$

where the symbol $\oplus$ denotes the summation $(J+K)$ modulo 2. The control bit remains unchanged, but its states determine the evolution of the target.

Combining these operations yields the discrete Fourier transformation (10.36)

$$\mathcal{U}_{\mathrm{FT}}\varphi_J^{(2)} = \frac{1}{2}\sum_{K=0}^{K=3}\exp(\mathrm{i}JK\pi/2)\varphi_K^{(2)}\,. \qquad (10.14)$$

n-Qubit Systems

A collection of n qubits is called a quantum register of size n. We assume that information is stored in the register in binary form. An n-register can store the numbers $J = 0, 1, \ldots, (2^n - 1)$. A quantum register of size 1 can store the numbers 0 and 1; of size 2, the numbers 0,1,2 and 3; etc.

The Hadamard gate, all the phase gates, and the controlled-NOT gate constitute a universal set of gates, although this set is not unique. Any transformation between the n states of a register may be constructed from them.

Both the control and the target can be generalized to become n-registers. In such a case, the symbol $\oplus$ in the first equation (10.13) denotes summation modulo 2^n. We may perform the operations

$$\mathcal{U}\varphi_J^{(n)}\varphi_K^{(n)} = \varphi_J^{(n)}\varphi_{K\oplus J}^{(n)}\,, \qquad \mathcal{U}_f\varphi_J^{(n)}\varphi_K^{(n)} = \varphi_J^{(n)}\varphi_{K\oplus f(J)}^{(n)}\,, \qquad (10.15)$$

where we define a function $f(J)$ mapping the number J to another number that may be stored by an n-register.

The discrete Fourier transformation (10.14) is written as

$$\mathcal{U}_{\mathrm{FT}}\varphi_J^{(n)} = 2^{-n/2}\sum_{K=0}^{K=2^n-1}\exp[\mathrm{i}JK\pi/2^{(n-1)}]\varphi_K^{(n)}\,. \qquad (10.16)$$

An important quantum algorithm permits the implementation of this transformation with a number of elementary one- and two-qubit gates that scales as n^2, while the classical Fourier transformation scales as $n2^n$.

10.4.3 Factorization

An algorithm due to Schor can be used to factorize the number N [68]. Let a be coprime with N (no common factors) and define the function

$$f_{aN}(J) \equiv (a^J, \mathrm{mod}\, N)\,. \qquad (10.17)$$

This function has at least two important properties:

- It is periodic. For instance, if $a = 2$, $N=15$, the successive values of the function f are $1, 2, 4, 8, 1, 2$, and so on. Thus, the period $P=4$.

10.4 Quantum Computation

- Provided that P is even, the greatest common divisors of the pairs $(a^{P/2}+1, \bmod N)$ and $(a^{P/2}-1, \bmod N)$ are factors of N. In the present example, they are 5 and 3, respectively.

Two thousand years ago, Euclid found a very efficient algorithm for calculating greatest common divisors. In contrast, the level of complexity in the calculation of the period using a classical computer is as large as any other factorization algorithm.

One of the promising features of quantum computation is the fact that we may transform a linear combination of states in a single run. Thus we apply the second transformation (10.15), with $K=0$, to the linear combination

$$\mathcal{U}_f \sum_{J=0}^{J=2^n-1} \frac{1}{\sqrt{2^n}} \varphi_J^{(n)} \varphi_0^{(n)} = \sum_{J=0}^{J=2^n-1} \frac{1}{\sqrt{2^n}} \varphi_J^{(n)} \varphi_{f_{aN}(J)}^{(n)} . \tag{10.18}$$

For $n=4$ and the previous example, the right-hand side of (10.18) takes the value

$$\frac{1}{4}\left[\varphi_0^{(4)} + \varphi_4^{(4)} + \varphi_8^{(4)} + \varphi_{12}^{(4)}\right]\varphi_1^{(4)} + \frac{1}{4}\left[\varphi_1^{(4)} + \varphi_5^{(4)} + \varphi_9^{(4)} + \varphi_{13}^{(4)}\right]\varphi_2^{(4)}$$
$$+ \frac{1}{4}\left[\varphi_2^{(4)} + \varphi_6^{(4)} + \varphi_{10}^{(4)} + \varphi_{14}^{(4)}\right]\varphi_4^{(4)} + \frac{1}{4}\left[\varphi_3^{(4)} + \varphi_7^{(4)} + \varphi_{11}^{(4)} + \varphi_{15}^{(4)}\right]\varphi_8^{(4)} . \tag{10.19}$$

Next we perform a measurement on the target register, obtaining one value f_i of the function f_{aN}. As a consequence, the control register collapses into a superposition of $\varphi_J^{(n)}$ with J yielding the same eigenvalue f_i. Quite generally, we may write $J = rP + q$, $r = 0, 1, 2, \ldots < 2^n/P$ and $q = 0, 1, \ldots, P-1$. This is still not very helpful for the determination of the period P, since we do not know the value of q. However, we may perform a discrete Fourier transformation, so that

$$\mathcal{U}_{\mathrm{FT}} \sum_{r=0}^{r<2^n/P} \Psi_{(rP+q)}^{(n)} = \sum_{K=0}^{K=2^n-1} \sum_{r=0}^{r<2^n/P} \exp\left[\frac{iK(rP+q)\pi}{2^{(n-1)}}\right] \varphi_K^{(n)}$$
$$= \sum_{r=0}^{r<2^n/P} c(f_i, rP)\varphi_{\nu=rP}^{(n)} . \tag{10.20}$$

The second step relies on the vanishing of the factor

$$\sum_{r=0}^{r<2^n/P} \exp\left[\frac{iK(rP)\pi}{2^{(n-1)}}\right] , \tag{10.21}$$

unless K is zero or an integer multiple of $2^n/P$, if P is an integer divisor of 2^n. Accordingly, the Fourier transformation of (10.19) yields

$$\left[c_{i0}\varphi_0^{(4)} + c_{i4}\varphi_4^{(4)} + c_{i8}\varphi_8^{(4)} + c_{i12}\varphi_{12}^{(4)} \right] \varphi_{f_i}^{(4)}, \tag{10.22}$$

where all subscripts are multiples of the period, and with coefficients given in Table 10.1.

Table 10.1. Values of amplitudes c_{ir} in the Fourier transform (10.22)

f_i	c_{i0}	c_{i4}	c_{i8}	c_{i12}
1	1	1	1	1
2	1	i	-1	$-i$
4	1	-1	1	-1
8	1	$-i$	-1	i

A variety of two-level quantum systems has been considered. Modern experimental techniques allow us to orient their spins and to implement the gates. However, the situation becomes drastically more complicated for operation with a large scale computer, combining many gates. The greatest problems lie in alteration of states due to decoherence, i.e., the unavoidable coupling with surrounding media (Sect. 11.1). Up to now the successes of quantum computation have been limited to the decomposition of the number 15 into its prime factors 3 and 5 [69].

It is true that we cannot ignore the example of the path traveled "from the Pascal machine to the Pentium processor." There exist new strategies for partially controlling the effects of decoherence. The fact that this problem is linked to defense and financial activities has undoubtedly contributed to intense endeavors on the subject. But we should bear in mind that the interest of quantum computing is not limited to those applications: the physics involved in experiments with entangled particles will help us to obtain a better understanding of the most fundamental aspects of quantum mechanics.

10.5* Bell States

A set of basis states for the two-spin system is given by the product states

$$\varphi_\downarrow\varphi_\downarrow, \quad \varphi_\downarrow\varphi_\uparrow, \quad \varphi_\uparrow\varphi_\downarrow, \quad \varphi_\uparrow\varphi_\uparrow, \tag{10.23}$$

and may therefore be represented by a four-component column vector. A general state is written as the superposition

$$\Psi_c = \frac{1}{\sqrt{|c_{\downarrow\downarrow}|^2 + |c_{\downarrow\uparrow}|^2 + |c_{\uparrow\downarrow}|^2 + |c_{\uparrow\uparrow}|^2}} \begin{pmatrix} c_{\downarrow\downarrow} \\ c_{\downarrow\uparrow} \\ c_{\uparrow\downarrow} \\ c_{\uparrow\uparrow} \end{pmatrix}. \tag{10.24}$$

Bell states constitute specific examples of entangled pairs

$$\varphi_{B_0} \equiv \frac{1}{\sqrt{2}}\begin{pmatrix}1\\0\\0\\1\end{pmatrix}, \quad \varphi_{B_1} \equiv \frac{1}{\sqrt{2}}\begin{pmatrix}1\\0\\0\\-1\end{pmatrix},$$

$$\varphi_{B_2} \equiv \frac{1}{\sqrt{2}}\begin{pmatrix}0\\1\\1\\0\end{pmatrix}, \quad \varphi_{B_3} \equiv \frac{1}{\sqrt{2}}\begin{pmatrix}0\\1\\-1\\0\end{pmatrix}. \tag{10.25}$$

- Since Bell states are orthonormal, any two-qubit state may be expressed as a linear combination of these states.
- The Bell states are eigenstates of the product operators $\hat{S}_z(1)\hat{S}_z(2)$ and $\hat{S}_x(1)\hat{S}_x(2)$, with eigenvalues given in Table 10.2.

Table 10.2. Eigenvalues of the product operators in units of $\hbar^2/4$

	φ_{B_0}	φ_{B_1}	φ_{B_2}	φ_{B_3}
$\hat{S}_z(1)\hat{S}_z(2)$	1	1	−1	−1
$\hat{S}_x(1)\hat{S}_x(2)$	1	−1	1	−1

- These product operators are included among the interactions in the controlling Hamiltonian (10.5), used to manipulate qubits.
- Successive introduction of these product interactions separates any two-qubit system into the 4 Bell channels, in a similar way as the interaction with the magnetic field splits the two channels associated with a single qubit (Sects. 2.4 and 5.2.1).

10.6* No-Cloning Theorem

This theorem states that the state of a particle cannot be copied onto another particle, while the original particle remains the same. This is completely different from what happens in classical mechanics. The following elementary proof is taken from reference [70]. Suppose we have two qubits in pure states

$$\varphi(1)\chi(2), \tag{10.26}$$

and that some unitary evolution effects the copying process

$$\varphi(1)\varphi(2) = \mathcal{U}\varphi(1)\chi(2). \tag{10.27}$$

Suppose now that this copying procedure also works for another state

$$\phi(1)\phi(2) = \mathcal{U}\phi(1)\chi(2) . \tag{10.28}$$

The scalar product between (10.27) and (10.28) yields

$$\langle\varphi|\phi\rangle^2 = \langle\chi\varphi|\mathcal{U}^+\mathcal{U}|\phi\chi\rangle = \langle\varphi|\phi\rangle . \tag{10.29}$$

Since this equation has two solutions, 0 and 1, either $\varphi = \phi$ or they are mutually orthogonal. Therefore, a general quantum cloning device is impossible.

Even if one allows non-unitary cloning devices, the cloning of non-orthogonal pure states remains impossible unless one is willing to tolerate a finite loss of fidelity in the cloned states. Similar conclusions also hold for general qubits.

10.7* Manipulations with Qubits

In this section we discuss the implementation of one- and two-qubit gates starting from the controlling Hamiltonian (10.5).

One Qubit

The Hadamard gate and the phase gate can be constructed by means of the following operations:

$$\mathcal{U}_\mathrm{H} = \mathcal{U}_z(\pi/2)\,\mathcal{U}_x(\pi/2)\,\mathcal{U}_z(\pi/2) = \frac{1}{\sqrt{2}}\begin{pmatrix} 1 & 1 \\ 1 & -1 \end{pmatrix}, \tag{10.30}$$

$$\mathcal{U}_\phi(\beta) = \mathcal{U}_z(-\beta) = \begin{pmatrix} 1 & 0 \\ 0 & \exp(\mathrm{i}\beta) \end{pmatrix}, \tag{10.31}$$

up to a phase. We obtain expressions (10.9) upon application of the matrices (10.30) and (10.31) to the column states (10.6).

Two Qubits

It is convenient to transcribe the matrix operators appearing for single qubits to the column basis (10.12). For instance,

$$\hat{S}_x^{(\mathrm{ctrl})} = \frac{\hbar}{2}\begin{pmatrix} 0 & 0 & 1 & 0 \\ 0 & 0 & 0 & 1 \\ 1 & 0 & 0 & 0 \\ 0 & 1 & 0 & 0 \end{pmatrix}, \quad \hat{S}_x^{(\mathrm{targ})} = \frac{\hbar}{2}\begin{pmatrix} 0 & 1 & 0 & 0 \\ 1 & 0 & 0 & 0 \\ 0 & 0 & 0 & 1 \\ 0 & 0 & 1 & 0 \end{pmatrix}. \tag{10.32}$$

For the sake of simplicity, we only illustrate the use of the interaction term (10.5) in the case of the SWAP gate. Consider the two-body term

$$\hat{H}_{xy} = J^{(\text{ctrl},\text{targ})}\left[\hat{S}_x^{(\text{ctrl})}\hat{S}_x^{(\text{targ})} + \hat{S}_y^{(\text{ctrl})}\hat{S}_y^{(\text{targ})}\right]$$

$$= \frac{\hbar^2}{2} J^{(\text{ctrl},\text{targ})} \begin{pmatrix} 1 & 0 & 0 & 0 \\ 0 & 0 & 1 & 0 \\ 0 & 1 & 0 & 0 \\ 0 & 0 & 0 & 1 \end{pmatrix}. \quad (10.33)$$

If applied during the time-interval τ, $\hat{H}_{xy}$ yields the transformation

$$\mathcal{U}_{\text{SWAP}}(\gamma) = \exp(i\hat{H}_{xy}\tau/\hbar) = \begin{pmatrix} 1 & 0 & 0 & 0 \\ 0 & \cos\gamma & i\sin\gamma & 0 \\ 0 & i\sin\gamma & \cos\gamma & 0 \\ 0 & 0 & 0 & 1 \end{pmatrix}, \quad (10.34)$$

where

$$\gamma = \frac{\hbar}{2} J^{(\text{ctrl},\text{targ})} \tau .$$

The $\mathcal{U}_{\text{SWAP}}(\pi/2)$ gate interchanges $\varphi_1^{(2)} \leftrightarrow \varphi_2^{(2)}$.

The controlled-NOT and the controlled-phase gates are expressed in matrix form by

$$\mathcal{U}_{\text{CNOT}} = \begin{pmatrix} 1 & 0 & 0 & 0 \\ 0 & 1 & 0 & 0 \\ 0 & 0 & 0 & 1 \\ 0 & 0 & 1 & 0 \end{pmatrix}, \quad \mathcal{U}_{\text{CB}}(\phi) = \begin{pmatrix} 1 & 0 & 0 & 0 \\ 0 & 1 & 0 & 0 \\ 0 & 0 & 1 & 0 \\ 0 & 0 & 0 & \exp(i\phi) \end{pmatrix}. \quad (10.35)$$

The construction of the controlled-NOT and controlled-phase gates starting from the Hamiltonian (10.5) is more involved than (10.34) and we omit it from this presentation.

For the two-qubit case, the Fourier transform is obtained through the following operations:

$$\mathcal{U}_{\text{FT}} = \mathcal{U}_{\text{SWAP}}(\pi/2)\, \mathcal{U}_{\text{H}}^{(\text{targ})} \mathcal{U}_{\text{CB}}(\pi/2) \mathcal{U}_{\text{H}}^{(\text{ctrl})} = \frac{1}{2}\begin{pmatrix} 1 & 1 & 1 & 1 \\ 1 & i & -1 & -i \\ 1 & -1 & 1 & -1 \\ 1 & -i & -1 & i \end{pmatrix}. \quad (10.36)$$

Problems

Problem 1. Under what conditions do the Bell states (Sect. 10.5*) remain invariant if single qubits are subject to the orthonormal transformation (2.12)?

Problem 2. Alice and Bob share three good qubits. What is the probability that Bob receives the correct message if there is an eavesdropper?

Problem 3. Find the generators of rotations that Bob has to perform in order to obtain the original qubit for each Bell state that Alice may have detected.

Problem 4. Express the Fourier transform of a single qubit in terms of universal gates.

Problem 5. Show that $\mathcal{U}_x(\pi)\mathcal{U}_z(\beta)\mathcal{U}_x(-\pi) = \mathcal{U}_z(-\beta)$.

Problem 6. Show that $\frac{\hbar}{2}\mathcal{U}_{\text{CNOT}}\hat{S}_x^{(\text{ctrl})}\mathcal{U}_{\text{CNOT}} = \hat{S}_x^{(\text{ctrl})}\hat{S}_x^{(\text{targ})}$.

11 Measurements and Alternative Interpretations of Quantum Mechanics. Decoherence

> *The problem of getting the interpretation proved to be rather more difficult than just working out the equations. [71]*

In this chapter we explore in greater depth some essential features of quantum mechanics, already discussed in Sects. 2.4 and 2.5. We now use the language associated with spin, a language the reader became familiar with in Chaps. 3 and 5. The time-dependence of spin states was discussed in Sect. 9.2. Filters have similar properties to those described in Sect. 2.4.

We also discuss the concept of decoherence, which allows for a novel interpretation of measurements (Sect. 11.2).

11.1 Measurements and Alternative Interpretations of Quantum Mechanics

Let us consider a magnetic field in the z-direction and assume that the spin points up in the x-direction at time t_0. We also include now the time-dependence of the states, as in (9.14). The state vector of the system is denoted by $\Psi(t_0) = \varphi_{x\uparrow}$ at $t = t_0$. Between t_0 and t_1, it evolves swiftly and deterministically to the state $\Psi(t_1)$, in accordance with the time-dependent Schrödinger equation. At t_1, we measure the spin projection along the positive x-axis by means of a filter. Using (9.14)

$$\Psi(t_1) = \varphi_{x\uparrow} \cos\frac{1}{2}\omega_L(t_1 - t_0) + i\varphi_{x\downarrow} \sin\frac{1}{2}\omega_L(t_1 - t_0). \qquad (11.1)$$

In the course of the measurement, each of the two possible results becomes correlated with different sets of macroscopic degrees of freedom, including those of the measurement and registration equipment. Let us assume that the eigenvalue $-\hbar/2$ is obtained at time t_1.

Before the first measurement, the linear combination (11.1) allows for the possible existence of interference effects. The measurement destroys such effects, even if we do not look at the results. What counts in the fact that in quantum mechanics any measurement involves a physical interaction between a microscopic system and a macroscopic apparatus that was devised by a humam being for a specific purpose [24].

After the measurement, the component $\varphi_{x\downarrow}$ constitutes the starting point for a new evolution of the system. A new measurement is carried out at time t_2. The expansion, analogous to (11.1), is written as

$$\Psi(t_2) = i\varphi_{x\uparrow} \sin \frac{1}{2}\omega_L(t_2 - t_1) + \varphi_{x\downarrow} \cos \frac{1}{2}\omega_L(t_2 - t_1) \,. \tag{11.2}$$

The probability amplitudes for obtaining the eigenvalues $\pm\hbar/2$ at time t_2 are given by the product

$$\langle \varphi_{x\uparrow} | \Psi \rangle_{t_2} = -\sin \frac{1}{2}\omega_L(t_2 - t_1) \sin \frac{1}{2}\omega_L(t_1 - t_0) \,,$$

$$\langle \varphi_{x\downarrow} | \Psi \rangle_{t_2} = i \cos \frac{1}{2}\omega_L(t_2 - t_1) \sin \frac{1}{2}\omega_L(t_1 - t_0) \,. \tag{11.3}$$

If no measurement had been performed at time t_1, the amplitudes would have been

$$\langle \varphi_{x\uparrow} | \Psi \rangle_{t_2} = \cos \frac{1}{2}\omega_L(t_2 - t_0) \,,$$

$$\langle \varphi_{x\downarrow} | \Psi \rangle_{t_2} = i \sin \frac{1}{2}\omega_L(t_2 - t_0) \,. \tag{11.4}$$

It is consistent to believe that the role of physics is to correlate the construction of the initial state $\Psi(t_0)$ with the results of the two measurements. In that case, (11.3) is all that is needed. In particular, questions about what happens in the intervals $t_1 - t_0$ and $t_2 - t_1$ become irrelevant. Bohr has warned us not to transpose our everyday experience to microscopic systems [72, 73]. This attitude is the opposite of that underlying the EPR criticism, where the 'objective reality' of the intermediate steps becomes of paramount importance [19].

The example of the two successive measurements also emphasizes a problem concerning quantum formalism, namely, the coexistence of two different time-evolutions of the state vector. Normally, it transforms swiftly in a deterministic way in accordance with the Schrödinger equation. However, if a measurement takes place, it changes suddenly and unpredictably.

When particles are detected, the atoms of the detector become ionized, producing first a few electrons, and then a cascade of electrons. The state vector should take these macroscopic effects into account. Because of the linearity of the Schrödinger evolution, there is no mechanism to stop the evolution and yield a single result for the measurement: the state reduction is beyond the scope of the Schrödinger evolution. Ultimately the evolution may involve the observer's brain, since the disappearance of macroscopic superpositions is attributed to the existence of the observer. Some extreme advocates of this interpretation have even argued that this mechanism may be linked to the property of consciousness in the human brain. Thus it is argued that quantum mechanics has an anthropocentric foundation, a concept which disappeared from science after the Middle Ages.

Although the Copenhagen interpretation has not been disproved, we should mention an alternative to state reduction, proposed by H. Everett III [74]. What the physical system does, together with the observer and the environment, is constantly split the state vector into all branches corresponding to each result of a measurement, without ever selecting one of them. Each component of the observer remains unaware of all the other components. In this sense, a quantum measurement never takes place. The illusion of the emergence of a single result appears as a consequence of the limitations of the human mind: the anthropocentric element creeps in again. In addition, the many-worlds interpretation creates many problems of its own, the main one being that, by definition, it cannot be disproved.

The many-worlds formulation has, unlike the traditional reduction representation, received support in literary works.[1]

There is a clash between the principle of superposition and the everyday classical reality in which a single outcome emerges, which is the main source of the unrest with quantum mechanics. Bohr's interpretation is based on the need for a classical apparatus to carry out measurements. Thus, there is a dividing line between the quantum and the classical, but without a clear identification of this borderline. The many-worlds interpretation aims to abolish the need for a border. Yet the lack of perception of the observed alternatives remains unexplained.

11.2 Decoherence

During the last 20 years, progress has been made on the interpretation problem discussed in Sect. 11.1, based on the realization that the linear, time-dependent, Schrödinger equation is only valid for closed systems.

All states are supposed to be quantum mechanical, including the much smaller set of classical states. The possible superpositions in Hilbert space are potentially expanded with the Schrödinger equation. The process of decoherence, i.e., the interaction between systems and environment, leads to the elimination of quantum superpositions and to the selection of a small subset of classical, pointer states [76].

In the following, we sketch how this may be accomplished for the measurement of a spin 1/2 system. Assume that the spin is, initially, in the state

$$\Psi^S(0) = c_\uparrow \varphi_\uparrow + c_\downarrow \varphi_\downarrow \,, \qquad |c_\uparrow|^2 + |c_\downarrow|^2 = 1 \,. \tag{11.5}$$

---

[1] "In all fictional works, each time that a man is confronted with several alternatives, he chooses one and eliminates the others; in the fiction of Ts'ui Pên, he chooses – simultaneously – all of them. He *creates*, in this way, diverse futures, diverse times which themselves also proliferate and fork .... In the work of Ts'ui Pên, all possible outcomes occur; each one is the point of departure of other forkings. Sometimes, the paths of this labyrinth converge: for instance, you arrive at this house, but in one of the possible pasts you are my enemy, in another my friend." [75].

Following John von Neumann, we consider a quantum apparatus $\mathcal{Z}$ with a Hilbert space spanned by the two states χ_1, χ_0 [77]. One can assume that the initial state of the binary spin–apparatus system is

$$\Psi^{\mathcal{S},\mathcal{Z}}(0) = \left(c_\uparrow \varphi_\uparrow + c_\downarrow \varphi_\downarrow\right) \chi_1 . \tag{11.6}$$

The entanglement of the composite system may be produced by means of the interaction represented by a controlled-NOT gate [see (10.35)]. Thus,

$$\Psi_t^{\mathcal{S},\mathcal{Z}} = c_\uparrow \varphi_\uparrow \chi_1 + c_\downarrow \varphi_\downarrow \chi_0 . \tag{11.7}$$

If the detector is in the state χ_1, the system is guaranteed to be found in the state $\varphi_\uparrow$, and vice versa. However, there is an ambiguity in a correlated state of the form (11.7), since we may rotate both the spin and the apparatus without changing $\Psi_t^{\mathcal{S},\mathcal{Z}}$ (see p. 154). The ambiguity may be superseded by introducing another system, the environment $\mathcal{E}$, which is also represented by the two quantum states $\varepsilon_1, \varepsilon_0$. Proceeding as in the former case,

$$\Psi^{\mathcal{S},\mathcal{Z},\mathcal{E}}(0) = \Psi_t^{\mathcal{S},\mathcal{Z}} \varepsilon_1 \longrightarrow \Psi_t^{\mathcal{S},\mathcal{Z},\mathcal{E}} = c_\uparrow \varphi_\uparrow \chi_1 \varepsilon_1 + c_\downarrow \varphi_\downarrow \chi_0 \varepsilon_0 . \tag{11.8}$$

In general, we cannot control the environment and we are limited to evaluating expectation values of observables belonging to the $(\mathcal{S}, \mathcal{Z})$ subsystems. In the state (11.8), any such expectation value is

$$\langle Q \rangle = |c_\uparrow|^2 \langle \varphi_\uparrow \chi_1 | Q | \varphi_\uparrow \chi_1 \rangle + |c_\downarrow|^2 \langle \varphi_\downarrow \chi_0 | Q | \varphi_\downarrow \chi_0 \rangle$$
$$+ 2\mathrm{Re}\left(c_\uparrow c_\downarrow^* \langle \varphi_\downarrow \chi_0 | Q | \varphi_\uparrow \chi_1 \rangle \langle \varepsilon_0 | \varepsilon_1 \rangle\right) . \tag{11.9}$$

The third term in this equation is responsible for introducing interference. Interaction with the environment has the effect of modulating this interference term, whose magnitude is reduced by a factor determined by the absolute value of the overlap $\langle \varepsilon_0 | \varepsilon_1 \rangle$. When this overlap is sufficiently small, one says that the environment induces decoherence in the system. Therefore, decoherence is effective when the states of the environment that become correlated with the two alternatives that are present in the state (11.8) are sufficiently different from each other. In this case, quantum interference effects become dynamically suppressed [76].

Not all states of the apparatus are equally susceptible to decoherence. In particular, states that diagonalize the interaction Hamiltonian between the system and the apparatus show minimal disturbance. They are the so-called pointer states of the apparatus and they are the only ones that are able to persist in a relatively stable way and become classically correlated with the system being measured.

Decoherence therefore explains why we do not see quantum superpositions in our everyday world: macroscopic objects are more difficult to keep isolated than microscopic objects. It also explains why spin $\uparrow$ and $\downarrow$ states are more easily preserved than their linear combinations through their interaction with the environment [78]: "Decoherence produces an effect that looks and smells like a collapse."

Decoherence is currently the subject of a great deal of research. Many questions have been clarified to a large extent in recent years. These include the rate of decoherence, the dynamical selection of the pointer states, the dissipation of energy into the environment, and many others.

12 A Brief History of Quantum Mechanics

12.1 Social Context

To continue the building analogy of Chap. 1, the theoretical foundations of physics were shaken at the beginning of the twentieth century. These tremors preceded those of society as a whole. The historian Eric Hobsbawm has written [79]:

> The decades from the outbreak of the First World War to the aftermath of the second, were an Age of Catastrophe for this society [...] shaken by two world wars, followed by two waves of global rebellion and revolution [...]. The huge colonial empires, built up before and during the Age of the Empire, were shaken, and crumbled to dust. A world economic crisis of unprecedented depth brought even the strongest capitalistic economies to their knees and seemed to reverse the creation of a single universal world economy, which had been so remarkable an achievement of nineteenth-century liberal capitalism. Even the USA, safe from war and revolution, seemed close to collapse. While the economy tottered, the institutions of liberal democracy virtually disappeared between 1917 and 1942 from all but a fringe of Europe and parts of North America and Australasia, as Fascism and its satellite authoritarian movements and regimes advanced.

Since quantum mechanics was developed for the most part in Northern and Central Europe (see Table 12.1), we will devote most of our attention here to the conditions prevailing at that time in Germany and Denmark.[1]

Hobsbawm's description applies particularly well to the case of Germany: while the Anglo-Saxon world and the wartime neutrals more or less succeeded in stabilizing their economies between 1922 and 1926, Germany was overwhelmed in 1923 with economic, political and spiritual crises. Hunger riots erupted everywhere, as the value of the mark plunged to 10^{-12} of its pre-1913 value. Additional difficulties arose from a repressed military putsch in North Germany, a separatist movement in the Rhineland, problems with France on the Rhur, and radical leftist tendencies in Saxony and Thuringia. In the East, Soviet Russia did not fare better.

---
[1] The sources [80–82] have been used extensively for this chapter.

A cultural movement against dogmatic rationalism gained ground in German society after the war. The most widely read book opposed causality to life, and assimilated physics into causality [83]. Moreover, a profound division along political, scientific and geographic lines started to grow in the German physics community. Right wing physicists were in general chauvinistic, ultraconservative, provincial, anti-Weimar, and anti-Semitic. They were interested in the results of experiments and dissociated themselves from quantum and relativity theory. On the opposite side, the Berlin physicists were labeled as liberal and theoretical. Note, however, that the German physicists of that time, with the possible exception of Einstein and Born, could only be labeled as liberal or progressive in comparison with Johannes Stark and Philipp Lenard. The adjective 'theoretical' (appearing also in the name of Bohr's Institute in Copenhagen) would be translated today as 'fundamental'. Although the main theoretical center was in Berlin, strong theoretical schools also flourished in Göttingen and Munich. The start of Nazi persecutions in the thirties and the exclusion of Jews from the first group had consequences in the world distribution of physicists devoted to the most fundamental aspects of physics.

After 1918, German physicists were excluded from international collaborations, and the lack of foreign currency made it almost impossible to purchase foreign journals and equipment. However, a new national organization, the Notgemeinschaft der Deutschen Wissenschaft, created in 1920 under the direction of Max von Laue and Max Planck, was instrumental in the provision of funds for scientific research. Atomic theorists in Berlin, Göttingen and Munich received sufficient funds to support the work of physicists like Heisenberg and Born. The foreign boycott was not observed by Scandinavia and the Netherlands: Bohr kept friendly relations with his German colleagues (see p. 180).

Denmark had been on the decline at least since 1864, when it was defeated by Prussia and Austria with the resultant loss of about one third of its territory. The years after the war represented a period of unprecedented turmoil in Denmark as well. For the first time in four hundred years, this country teetered on the brink of revolution, although of a kind that was different from those experienced in neighboring countries. Disputes over the shift of the border with Germany, social struggles between town and country, fights for extensive reforms in employment conditions; all these difficulties added to the loss of wartime markets, and to trade deficits and inflation. In spite of such hardships, Bohr's new institute was inaugurated in 1921.

12.2 Old Quantum Theory ($1900 \leq t \leq 1925$)

The first determination of the mean lifetime of a radioactive substance (Rn^{220}) was made by Ernest Rutherford [84]. He noted that the decrease in the number N of atoms with time t is well described by the law

$$N(t) = N(0)\exp(-\lambda t) , \qquad (12.1)$$

a relation implying that identical atoms may decay at different times. This is impossible to explain using classical arguments. In 1916, Einstein was the first to realize that (12.1) can only be understood within the context of quantum theory.

In the late 1890s, crucial measurements of the spectral distribution of black-body radiation were performed in Berlin. Planck made an extraordinarily successful guess, Planck's radiation law, reproducing the spectral distribution as a function of frequency and temperature. Two months later he justified his law by means of the quantum hypothesis: the oscillators of a black body have energy

$$\epsilon = h\nu , \qquad (12.2)$$

where Planck's constant h was introduced and ν is the frequency [2].

It was Einstein who realized that radiation itself had a discrete structure (12.2), a hypothesis that went far beyond Planck's suggestion, and that was resisted by Planck for a long time. On the basis of (12.2), Einstein interpreted the photoelectric effect as the total transfer of the photon energy to the electron, whose energy W is given by

$$W = h\nu - P , \qquad (12.3)$$

where P is the energy loss suffered by the electron before it reaches the surface of the metal. This relation was only confirmed experimentally in 1916. (However, none of the experimentalists concluded in favor of Einstein's "bold, not to say reckless hypothesis", as Robert Millikan called it in 1916.)

Evidence of the wave nature of light had been abundant. Yet the discovery of the photoelectric effect, and later on of the Compton effect, provided evidence for a particle-like behavior of light. Even in 1909, Einstein concluded: "It is my opinion that the next phase of theoretical physics will bring us a theory of light which can be interpreted as a kind of fusion of the wave and the emission theory." From 1906 to 1911, quantum theory was Einstein's main concern (more than the theory of relativity). He made theoretical contributions to the specific heat of solids, the momentum of the light particle, and so on.

In 1911, the work of Rutherford's young colleagues Hans Geiger and Ernest Mardsen showed conclusively that a hydrogen atom consists of one electron outside the positively charged nucleus, where almost all the mass is concentrated [6]. At that time, electrons were supposed to be just particles. (Electron wave behaviour was experimentally verified in [5].) What happened inside the atom? The Balmer formula fitting the frequencies of the discrete hydrogen spectrum dated from 1885 [7]:

$$\nu_n = cR_{\mathrm{H}}\left(\frac{1}{4} - \frac{1}{n^2}\right) , \qquad n = 3, 4, \ldots , \qquad (12.4)$$

where R_H is the Rydberg constant. Unlike Planck's case, no significant progress was made in understanding Balmer's formula until the Bohr papers were published [8].

Like Einstein in 1905, Bohr was aware that his model was in conflict with classical physics. After the work of Newton and Maxwell and all of their followers, it required great boldness to assert his first postulate that an atom displays stationary states that do not radiate. Furthermore, this assumption also implied a renunciation of causality because of the absence of any directive for the transitions between stationary states which were accompanied by monochromatic radiation that satisfied the Balmer series. The derivation of the Rydberg constant as a function of the mass and the charge of an electron and of Planck's constant, and the correct helium ion/hydrogen ratio to five significant figures, commanded the attention of the physics community. Bohr made use of the correspondence principle, in which, for instance, he used the classical connection between the frequencies of the electron rotation and the emitted light in classical theory. This principle constituted the main link between classical and quantum theory. (See Fig. 4.2 as an illustration of the survival of the correspondence principle in quantum mechanics.)

By means of his precise determination of X-ray energies, Henry Moseley gave further support to the Bohr model both through the assignment of a Z value to all known elements and the prediction of the still missing elements [85]. James Frank and Heinrich Hertz further confirmed the model by using the impact of electrons on atoms to excite their atomic spectrum [86]. The Bohr model appeared to work, in spite of its assumptions. In order to joke about the situation with the old quantum theory, Bohr used to tell the story of a visitor to his country home who noticed a horseshoe hanging over the entrance door. Puzzled, he asked Bohr if he really believed that this brought luck. The answer was: "Of course not, but I am told it works even if you don't believe in it." [82].

Bohr developed his model during a post-doctoral stay at Rutherford's laboratory (Manchester, UK). He was appointed professor of physics at the University of Copenhagen in 1916 and the Universitetets Institut for Teoretisk Fysik (today, Niels Bohr Institutet) was inaugurated in 1921. Unlike Einstein and Dirac, Bohr seldom worked alone. His first collaborator was Hendrik Kramers (Netherlands), followed by Oscar Klein (Sweden). During the 1920s, there were 63 visitors to the institute who stayed more than one month, including 10 Nobel Laureates. The flow of foreign visitors lasted throughout Bohr's life: he became both an inspiration and a father figure.

Arnold Sommerfeld established an important school in Munich. In 1914, it was found that every line predicted by the Balmer formula is a narrowly split set of lines. By taking into account the influence of relativity theory, Sommerfeld showed that the orbits of the electrons are approximate ellipses displaying a parahelion precession [87]. Sommerfeld's work was also one of the first attempts to unite the quantum and relativity theories, a synthesis still not completely achieved.

In Göttingen, Born did not turn his attention to atomic theory until around 1921. Heisenberg and Jordan were his assistants.

In 1924 Pauli had published fifteen papers on topics ranging from relativity to the old quantum theory, the first one before entering the University of Munich. In 1922, he went for a year to Copenhagen. Later he assumed a position at Hamburg. He made the assumption of two-valuedness for electrons and stated the exclusion principle [37] (Sect. 7.1) so important for understanding the properties of atoms, conductors, nuclei, baryons, etc.

The crucial experimental results of Stern and Gerlach were known in 1921 [30]. A proposal concerning spin was made[2] by two Dutch students from the University of Leiden, Uhlenbeck and Goudsmit, who also suggested the existence of $m_s = \pm 1/2$ as a fourth quantum number [31] (Sect. 5.2.2). They explained the anomalous Zeeman effect by taking into account the factor of two appearing in (5.25). The explanation of this factor was supplied in [88]. After receiving objections from Henrik Lorentz, Uhlenbeck and Goudsmit considered withdrawing their paper, but it was too late. (Their superior, Paul Ehrenfest, also argued that the authors were young enough to be able to afford some stupidity.) The two-component spin formalism (5.24) was introduced by Pauli in 1927 [32].

Satyendra Bose's derivation of Planck's law using symmetric states was translated and submitted for publication by Einstein [39]. The same year Einstein applied Bose's ideas to an ideal gas of particles [44] (see Sect. 7.7* on Bose–Einstein condensation). Dirac developed quantum statistics for antisymmetric wave functions [40], and this was illustrated by Enrico Fermi using helium [41].

However, until 1925, there were almost as many setbacks as successes in the application of the model. For instance, the spectrum of He proved to be intractable, in spite of heroic efforts by Kramers, Heisenberg and others.

12.3 Quantum Mechanics ($1925 \leq t \leq 1928$)

Periodically, Bohr used to gather his former assistants together at his institute in Copenhagen. In 1925, the ongoing crisis in quantum mechanics was examined by Bohr, Kramers, Heisenberg and Pauli. A few months later, back at Göttingen, Heisenberg found a way out of the impasse [9]. He succeeded in

---

[2] The combination of the Pauli principle and of spinning electrons prompted Ralph Kronig to propose the idea of half-integer spin. However, he was dissuaded from publishing it by Pauli and others, on the grounds that models for electrons carrying an intrinsic angular momentum $\hbar/2$ either required the periphery of the electron to rotate with a velocity much larger than the velocity of light c, or the electron radius to be much larger than the classical value e^2/Mc.

formulating the theory in terms of observable quantities [for instance, doing away with the concepts of orbits and trajectories (see Sect. 2.1)]. Heisenberg found a correspondence between the coordinate $x(t)$ and the double array x_{nm} (n,m labeling quantum states). $x_{nm}(t)$ was interpreted as a sort of transition coordinate, and hence an allowed observable. In order to represent $x^2(t)$, he made the crucial assumption that $x^2 = \sum_p x_{np} x_{pm}$. Heisenberg solved the simple but non-trivial problem of the harmonic oscillator by showing that the Hamiltonian is $H_{mn} = E_n \delta_{nm}$, where the E_n reproduce the correct eigenvalues (Sect. 3.2).

Born and Jordan realized that the arrays (x_{nm}) were matrices and arrived at the fundamental commutation relation (2.8) in its matrix form (3.50) [89]. Born, Heisenberg and Jordan wrote a comprehensive text on quantum mechanics, which included unitary transformations, perturbation theory, the treatment of degenerate systems and commutation relations for the angular momentum operators [10].

Many of these results were also obtained by Dirac [11], who introduced the idea that physical quantities are represented by operators (of which Heisenberg's matrices are just one representation), the idea of representing physical states by rays in abstract Hilbert spaces, and the connection between the commutator of two operators with the classical Poisson bracket.

Fig. 12.1. The 1930 Copenhagen Conference. In the front row: Klein, Bohr, Heisenberg, Pauli, Gamow, Landau and Kramers. (Reproduced with permission from the Niels Bohr Archive, Copenhagen)

12.3 Quantum Mechanics ($1925 \leq t \leq 1928$)

In 1926, Pauli and Dirac independently reproduced the results for the hydrogen atom of old quantum theory using the new matrix mechanics [90, 91].

Zurich-based Schrödinger did not belong to the Copenhagen–Göttingen–Munich tradition. In 1925 he came across de Broglie's suggestion [28] that the wave–corpuscle duality should also be extended to material particles, satisfying the momentum–frequency relation (4.32). This relation is reproduced if the momentum and the energy are replaced by the differential operators (4.4) and (9.5) and if such operators act on the plane wave solutions (4.30) and (9.9). Upon making the same substitution in a general Hamiltonian, Schrödinger derived the time-independent and the time-dependent equations that bear his name [12]. Quantization was obtained through the requirement that the wave function should be single-valued [as in (5.46)]. Schrödinger presented his derivation as a step towards a continuous theory, the integers (quantum numbers) originating in the same way as the number of nodes in a classical vibrating string. Schrödinger's formulation gained rapid acceptance, both because of the answers that it produced and because it was built from mathematical tools that were familiar to the theoretical physicists of the time. Schrödinger hoped that quantum mechanics would become another branch of classical physics: waves would be the only reality, particles being produced by means of wave packets. This expectation turned out to be wrong.

In 1926, Schrödinger also proved that the matrix and the differential formulation are equivalent. Since physicists understood how to transcribe the language of wave mechanics into matrix mechanics, both of them were referred to as quantum mechanics.

The probability interpretation of $|\Psi(x,t)|^2$ soon became apparent (Sect. 4.1.1), and it is usually considered part of the Copenhagen interpretation. However, Born was the first to write it explicitly [10]. In his paper on collision theory, he also stated that $|c_i|^2$ (2.3) was the probability of finding the system in the state i. He emphasized that quantum mechanics does not answer the question: what is the state after a collision? Rather it tells us how probable a given effect of the collision is. Determinism in the atomic world was thereby explicitly abandoned.

In 1926 Heisenberg was able to account for the He problem using the Schrödinger equation plus the Pauli principle plus spin (Sect. 8.3) [92].

The relativistic generalization of the Schrödinger time-dependent, two-component spin formalism (5.24) encountered some difficulties. In 1928, Dirac produced an equation, linear in both the coordinates and time derivatives, with the properties that:

- it is Lorentz invariant,
- it satisfies a continuity equation (4.16) with positive density ρ (which previous attempts at relativistic generalization had failed to do)
- it encompasses spin from the start,

- it reproduces the results of the Sommerfeld model for the H atom, which were more accurate than the predictions of (new) quantum mechanics [13].

The price that Dirac had to pay for this most beautiful product of twentieth century mathematical physics was that it turned out to be a four-component rather than a two-component theory (5.24). Its interpretation including the additional two components is beyond the scope of this exposition.

A few comments are in order:

- Quantum mechanics and its traditional interpretation were developed over only a few years (1925–1928). The rate of publication in this period was such that many physicists complained about the impossibility of keeping up to date. Moreover, communication delays certainly hampered the ability of non-European physicists to contribute.
- Unlike previous scientific cornerstones, quantum mechanics was the result of the coherent effort of a group of people, mostly in Northern and Central Europe. Table 12.1 shows the number of papers written in each country during the period of major activity in the creation of quantum mechanics. It reflects both the predominance of Germany and the number of visitors, especially in the case of Denmark. It also reminds us that the scientific center of gravity was only transferred to the other side of the Atlantic after the Second World War (1939–1945).
- The Bohr Festspiele took place at Göttingen in 1922. After Bohr's speech, the 20 year old Heisenberg stood up and raised objections to Bohr's calculations. During a walk in the mountains that same afternoon, Bohr invited Heisenberg to become his assistant in Copenhagen. This anecdote points out the extreme youth (and self-confidence) of most of the contributors to quantum mechanics. In 1925 Dirac was 23 years old, Heisenberg was 24, Jordan 22, and Pauli 25. The 'elders' were Bohr (40), Born (43), and Schrödinger (38).

Table 12.1. Publications in quantum mechanics. July 1925 to March 1927 [80]

Country	Papers written	Country	Papers written
Germany	54	France	12
USA	26	USSR	11
Switzerland	21	Netherlands	5
Britain	18	Sweden	5
Denmark	17	Others	7

12.4 Philosophical Aspects

12.4.1 Complementarity Principle

Neither the Heisenberg nor the Schrödinger formulations improved the contemporary understanding of wave–particle duality. In 1927, Heisenberg answered the question: can quantum mechanics represent the fact that an electron finds itself approximately in a given place and that it moves approximately with a given velocity, and can we make these approximations so close that they do not cause mathematical difficulties? [93]. The answer was given in terms of the uncertainty relations (2.28) and (9.34) (see the last paragraph of Sect. 2.6). Heisenberg's paper was the beginning of the discussion of the measurement problem in quantum mechanics (Chap. 11), on which so many volumes have been written.

As most theoretical physicists would have done, Heisenberg derived his uncertainty principle from the (matrix) formalism (Sect. 2.6). Bohr had the opposite attitude. While being duly impressed by the existence of at least two formulations predicting correct quantum results, he insisted on first understanding the philosophical implications, rather than the formalisms. His main tools consisted of words, which he struggled continually to define precisely. Bohr pointed out that theories – even quantum theories – were checked by readings from classical instruments. Therefore, all the evidence has to be expressed within classical language, in which the mutually exclusive terms 'particle' and 'wave' are well defined. Either picture may be applied in experimental situations, but the other is then inapplicable. The idea of complementarity is that a full understanding of this microscopic world comes only from the possibility of applying both pictures; neither is complete in itself. Both must be present, but when one is applied, the other is excluded. These ideas were stated for the first time at the Como Conference, September 1927 [72]. Bohr continued to reformulate the presentation of the concept of complementarity throughout the rest of his life [94].

12.4.2 Bohr and Einstein

The famous discussions between Bohr and Einstein started in Brussels at the Solvay Conference of 1927, where Einstein expressed his concern over the extent to which the causal account in space and time had been abandoned in quantum mechanics. The discussions centered on whether a fuller description of phenomena could be obtained through the detailed balance of energy and momentum in individual processes. It was on this occasion that Einstein asked whether God had recourse to playing dice, to which Bohr replied by calling for great caution in ascribing attributes to Providence in everyday language.

At the next Solvay meeting in 1930, Einstein claimed that a control of energy and time could be achieved using relativity theory. He proposed the

Fig. 12.2. Einstein and Bohr leaving the Solvay meeting of 1930. (Reproduced with permission from the Niels Bohr Archive, Copenhagen)

device represented in Fig. 12.3, consisting of a box with a hole on a wall and a clock inside, such that a single photon might be released at a known moment. Moreover, it would be possible to measure the energy of the photon with any prescribed accuracy by weighing the box before and after the event, and make use of the relativity equation $E = mc^2$. Bewilderment among quantum physicists lasted until the next day, when Bohr came up with an answer based on general relativity: since the rate of the clock depends on its position in a gravitational field, the lack of precision in the box displacement generates an uncertainty Δt in the determination of time, while the indeterminacy in the energy ΔE is obtained through the position–momentum relation (2.28). The product $\Delta t \Delta E$ satisfies the Heisenberg time–energy uncertainty relation.

In 1935, Einstein, Podolsky and Rosen presented a profound argument pointing to the incompleteness of quantum mechanics [19]. They considered a system consisting of two entangled, but spatially separated particles, for which they assumed three premises:

- quantum mechanical predictions for this system are correct,
- the criterion for reality enunciated at the beginning of Sect. 2.1,
- the non-existence of actions at a distance.

Einstein, Podolsky and Rosen inferred the value of both the position and momentum of one of the particles by manipulating the other. They thus reached the conclusion that a quantum mechanical description using the wave func-

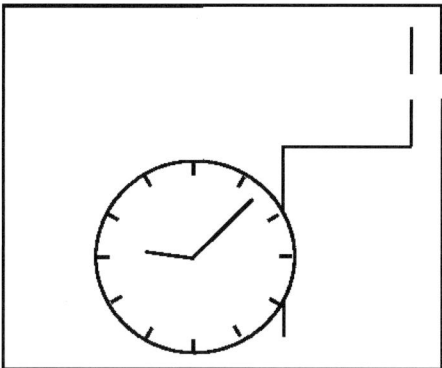

Fig. 12.3. Sketch of the thought experiment proposed by Einstein in order to reject the time–energy uncertainty relation. (Reproduced with permission from the Niels Bohr Archive, Copenhagen)

tion is incomplete. The quantum state only appeared to be indeterminate because some parameters characterizing the system were unmeasurable. These hidden variables should be incorporated into the description of physical systems, since they determine the outcome of experiments.

Most of the physics community rejected the EPR conclusion because of Bohr's reply, which was based on an analysis of the EPR definition of 'physical reality'. According to Bohr, these two terms can only be used in microphysics if the (mutually exclusive) experimental setups are specified [73].

Probably the best description of Einstein's and Bohr's respective positions is stated in Bohr's presentation on the occasion of Einstein's seventieth birthday [95]. In his answer in the same volume, Einstein stated [96]: "To believe that it [quantum mechanics] should offer an exhaustive description of individual phenomena is logically possible without contradiction; but [...] I cannot forego the search for a more complete description."

12.4.3 Recent History

Rather than dwell on philosophical interpretations of equations, most physicists proceeded to carry out many exciting applications of quantum mechanics [78]:

> This approach proved stunningly successful. Quantum mechanics was instrumental in predicting antimatter, understanding radioactivity (leading to nuclear power), accounting for the behavior of materials such as semiconductors, explaining superconductivity, and describing interactions such as those between light and matter (leading to the invention of the laser) and of radio waves and nuclei (leading to magnetic resonance imaging). Many successes of quantum mechanics

involve its extension, quantum field theory, which forms the foundations of elementary particle physics ....

On the other hand, the controversy over the EPR experiment did not die away, but the important issue turned out to be the locality of quantum mechanics, not its completeness. In 1952, David Bohm was able to reproduce the quantum mechanical predictions for systems of the EPR-type using a deterministic hidden-variable formulation, including nonlocal interactions between spatially separated particles [97].

In 1964, John Bell proved that nonlocality is an inherent characteristic of hidden variable theories reproducing the results of quantum mechanics [98]. Moreover, he showed that the correlations between measured properties of any classical, deterministic, local two-particle theory would obey a mathematical inequality. On the contrary, the same measurements would violate such an inequality if the two particles were in an entangled quantum state. Three-particle entangled states are predicted to display even more spectacular contradictions.

Following instrumental improvements in the production of entangled photons, Bell's contribution was followed by many proposals of possible experiments. They culminated in the results of [99] and [100], which are compatible with predictions of quantum mechanics and in conclusive disagreement with the results of local hidden-variable theories.

However, one may not conclude that these tests have *proven* the validity of quantum mechanics. It is worthwhile to remember that experiments can only prove that a theory is incorrect, if their results contradict the predictions of the theory. The validity of quantum mechanics should be further tested by checking consequences of modifications of quantum principles [such as the linearity of the Schrödinger equations (9.5)] against experiment.

13 Solutions to Problems and Physical Constants

13.1 Solutions to Problems

Chapter 2

Problem 1. (1) $\Psi = \dfrac{c_1}{\sqrt{|c_1|^2 + |c_2|^2}}\Psi_1 + \dfrac{c_2}{\sqrt{|c_1|^2 + |c_2|^2}}\Psi_2$.

(2) $\dfrac{|c_1|^2}{|c_1|^2 + |c_2|^2}$.

Problem 2. (1) $\Psi_3 = -\dfrac{c}{\sqrt{1 - |c|^2}}\Psi_1 + \dfrac{1}{\sqrt{1 - |c|^2}}\Psi_2$.

(2) $\Psi = \dfrac{(c_1 + cc_2)\Psi_1 + c_2\sqrt{1 - |c|^2}\Psi_3}{\sqrt{|c_1|^2 + |c_2|^2 + c_1 c^* c_2^* + c_1^* c c_2}}$.

Problem 5. $\mathrm{i}\dfrac{\hat{p}}{M}$.

Problem 6. $-\mathrm{i}n\hat{p}^{n-1}$.

Problem 8. (1) $\hat{R}\Psi = \Psi$. (2) $\hat{R}\Psi = 0$.

Problem 9. $\dfrac{\langle i|p|j\rangle}{\langle i|x|j\rangle} = \dfrac{\mathrm{i}M(E_i - E_j)}{\hbar}$.

Problem 10. (1) $\langle \Psi|Q|\Psi\rangle = \dfrac{11q}{6}$. (2) $P(q) = \dfrac{1}{6}$, $P(2q) = \dfrac{5}{6}$.

(3) $\dfrac{1}{\sqrt{5}}\Psi_2 + \dfrac{2}{\sqrt{5}}\Psi_3$.

Problem 11. $\Delta x \approx 10^{-19}$ m, $\Delta v \approx 10^{-19}$ m/s.

Problem 12. (1) $\Delta E_\mathrm{H}/\Delta E = O(10^{-25})$. (2) $x = O(10^{10})$ m.

Chapter 3

Problem 1. (1) $0, \pm\sqrt{2}$. (2) $\varphi_{\pm\sqrt{2}} = \dfrac{1}{2}\begin{pmatrix}1\\0\\0\end{pmatrix} \pm \dfrac{1}{\sqrt{2}}\begin{pmatrix}0\\1\\0\end{pmatrix} + \dfrac{1}{2}\begin{pmatrix}0\\0\\1\end{pmatrix}$,

$$\varphi_0 = \frac{1}{\sqrt{2}}\begin{pmatrix}1\\0\\0\end{pmatrix} - \frac{1}{\sqrt{2}}\begin{pmatrix}0\\0\\1\end{pmatrix}, \quad \mathcal{U} = \frac{1}{2}\begin{pmatrix}1 & \sqrt{2} & 1\\ \sqrt{2} & 0 & -\sqrt{2}\\ 1 & -\sqrt{2} & 1\end{pmatrix}.$$

Problem 2. (1) $\Delta_\pm = \pm\sqrt{a^2+c^2}$. (2) $\Delta_\pm = \pm|a|\left(1+\dfrac{c^2}{2a^2}+\cdots\right)$.

(3) $\Delta_\pm = \pm|c|\left(1+\dfrac{a^2}{2c^2}+\cdots\right)$.

Problem 3. $\langle 1|2\rangle = \langle 1|3\rangle = \langle 2|4\rangle = \langle 3|4\rangle = 0$.

Problem 4. (1) $\Delta_Q = (0.5, 0.5, -1)$, $\Delta_R = (0.5, -0.5, 1)$.

(2) $[\hat{Q}, \hat{R}] = 0$. (3) $\begin{pmatrix}1/\sqrt{2}\\ 1/\sqrt{2}\\ 0\end{pmatrix}$, $\begin{pmatrix}1/\sqrt{2}\\ -1/\sqrt{2}\\ 0\end{pmatrix}$, $\begin{pmatrix}0\\0\\1\end{pmatrix}$.

Problem 5. (1) $\pm\dfrac{\hbar}{2}$. (3) $\varphi_{\beta\uparrow} = \cos\dfrac{\beta}{2}\begin{pmatrix}1\\0\end{pmatrix} + \sin\dfrac{\beta}{2}\begin{pmatrix}0\\1\end{pmatrix}$,

$\varphi_{\beta\downarrow} = -\sin\dfrac{\beta}{2}\begin{pmatrix}1\\0\end{pmatrix} + \cos\dfrac{\beta}{2}\begin{pmatrix}0\\1\end{pmatrix}$.

Problem 7. (1) $E = V_0 + \dfrac{7\hbar}{2}\sqrt{\dfrac{c}{M}}$. (2) $E = -\dfrac{b^2}{2c} + \hbar\sqrt{\dfrac{c}{M}}\left(n+\dfrac{1}{2}\right)$.

Problem 8. (1) $x_c = \sqrt{\hbar}/(Mc)^{1/4}$. (2) $3\hbar\omega$.

Problem 9. (1) $\dfrac{2M\omega}{\hbar}\langle n+2|x^2|n\rangle = -\dfrac{2}{\hbar M\omega}\langle n+2|p^2|n\rangle$
$= \sqrt{(n+1)(n+2)}$, and 0 otherwise. (3) $\dfrac{\langle n|K|n\rangle}{\langle n|V|n\rangle} = -\dfrac{\langle n\pm 2|K|n\rangle}{\langle n\pm 2|V|n\rangle} = 1$.

Problem 10. Zero.

Problem 11. (1) $\Psi = \dfrac{1}{\sqrt{2}}\varphi_0 + \dfrac{1}{\sqrt{2}}\varphi_1$. (2) $\langle\Psi|x|\Psi\rangle = x_c$, $\langle\Psi|p|\Psi\rangle = \langle\Psi|\Pi|\Psi\rangle = 0$.

Chapter 4

Problem 3. (1) $\varphi_n = \sqrt{\dfrac{2}{a}}\sin(k_n x)$ $(0 \le x \le a)$ and $\varphi_n = 0$ otherwise. $k_n = n\pi/a$, $E_n = \hbar^2 k_n^2/2M$. (3) No.

Problem 4. $E \approx (\hbar\Delta p)^2/2M \ge \hbar^2/8Ma^2$.

Problem 5. $i\kappa \dfrac{1+\exp(-\kappa a)}{1-\exp(\kappa a)} = k \dfrac{1+\exp(ika)}{1-\exp(-ika)}$, $\kappa = \sqrt{2M(V_0-E)}/\hbar$,
$k = \sqrt{2ME}/\hbar$.

Problem 6. $\sum_k E_k - \dfrac{a}{2\pi} \int E_k \mathrm{d}k \approx E_{k_{\max}} = \hbar^2 k_{\max}^2/2M$.

Problem 7. $R = 0.030$, $T = 0.97$.

Problem 8. (1) $x_\mathrm{d} \approx 1/\kappa = 1.13$ Å.   (2) $T = 1.7 \times 10^{-15}$.

Problem 9. $\lim\limits_{\kappa a \ll 1} T = \dfrac{2E/V_0}{2E/V_0 + MV_0 a^2/\hbar^2}$,

$\lim\limits_{\kappa a \gg 1} T = \dfrac{16E}{V_0}\left(1-\dfrac{E}{V_0}\right)\exp\left[-\dfrac{a}{\hbar}\sqrt{2M(V_0-E)}\right]$.

Problem 10. (1) $-\cot\dfrac{ka}{2} = \dfrac{\kappa}{k}$.

Problem 11. 1 eps, 1 eps + 1 ops, 2 eps + 1 ops, 2 eps + 2 ops.

Problem 12. (2) The lattice exerts forces on the electron.
(3) $\langle k|p|k\rangle = \hbar k \int |u_k|^2 \mathrm{d}x - i\hbar \int u_k^* \dfrac{\mathrm{d}u_k}{\mathrm{d}x}\mathrm{d}x$.

Chapter 5

Problem 1. $O\left(10^{31}\right)$.

Problem 2. $\dfrac{\hbar}{\sqrt{2}}\begin{pmatrix} 0 & 1 & 0 \\ 1 & 0 & 1 \\ 0 & 1 & 0 \end{pmatrix} \longrightarrow \hbar \begin{pmatrix} 1 & 0 & 0 \\ 0 & 0 & 0 \\ 0 & 0 & -1 \end{pmatrix}$.

Problem 4. $i\hbar \boldsymbol{J}$.

Problem 5. (1) $\langle 00|Y_{20}|00\rangle = \langle 11|Y_{21}|21\rangle = \langle 00|Y_{11}|11\rangle = \langle 00|\Pi|10\rangle = 0$.
(2) $\langle 10|Y_{20}|10\rangle = 0.25$, $\langle 00|Y_{11}|1(-1)\rangle = -0.28$, $\langle 00|\Pi|00\rangle = -\langle 11|\Pi|11\rangle = 1$.

Problem 7. (1) $\varphi_{s_x=\pm\frac{1}{2}} = \dfrac{1}{\sqrt{2}}\begin{pmatrix} 1 \\ \pm 1 \end{pmatrix}$, $\varphi_{s_y=\pm\frac{1}{2}} = \dfrac{1}{\sqrt{2}}\begin{pmatrix} 1 \\ \pm i \end{pmatrix}$.
(2) $\pm\dfrac{\hbar}{2}$, $\dfrac{1}{2}$.   (3) $\hat{S}_x = \dfrac{\hbar}{2}\begin{pmatrix} 0 & i \\ -i & 0 \end{pmatrix}$.
(4) $\varphi_{s_x=\pm\frac{1}{2}} = \dfrac{1\pm i}{2}\varphi_{s_y=\frac{1}{2}} + \dfrac{1\mp i}{2}\varphi_{s_y=-\frac{1}{2}}$.

Problem 8. (1) $\frac{1}{2}(a+b)^2$. (2) $\frac{1}{2}(a^2+b^2)$. (3) a^2 .

Problem 9. (1) $\frac{\hbar}{2}$, $\cos^2\frac{\beta}{2}$, $-\frac{\hbar}{2}$, $\sin^2\frac{\beta}{2}$. (2) $\frac{\hbar}{2}\cos\beta$.

Problem 10. (1) $\varphi_{\frac{3}{2}\frac{1}{2}} = -\sqrt{\frac{2}{5}}Y_{20}\begin{pmatrix}1\\0\end{pmatrix} + \sqrt{\frac{3}{5}}Y_{21}\begin{pmatrix}0\\1\end{pmatrix}$,

$\varphi_{\frac{5}{2}\frac{1}{2}} = \sqrt{\frac{3}{5}}Y_{20}\begin{pmatrix}1\\0\end{pmatrix} + \sqrt{\frac{2}{5}}Y_{21}\begin{pmatrix}0\\1\end{pmatrix}$. (3) $Y_{ll}\begin{pmatrix}1\\0\end{pmatrix}$, 1 .

Problem 11. Equation (7.12).

Problem 12. $\sum_{m_1 m_2} c(j_1 m_1, j_2 m_2, jm)c(j_1 m_1, j_2 m_2, j'm') = \delta_{jj'}\delta_{mm'}$,
$\sum_{jm} c(j_1 m_1, j_2 m_2, jm)c(j_1 m'_1, j_2 m'_2, jm) = \delta_{m_1 m'_1}\delta_{m_2 m'_2}$.

Chapter 6

Problem 1. 2.5×10^{-3} eV.

Problem 2. (1) $1s_{\frac{1}{2}}$, $2s_{\frac{1}{2}}$, $2p_{\frac{1}{2}}$, $2p_{\frac{3}{2}}$, $3s_{\frac{1}{2}}$, $3p_{\frac{1}{2}}$, $3p_{\frac{3}{2}}$, $3d_{\frac{3}{2}}$, $3d_{\frac{5}{2}}$.
(2) $0s_{\frac{1}{2}}$, $1p_{\frac{1}{2}}$, $1p_{\frac{3}{2}}$, $2s_{\frac{1}{2}}$, $2d_{\frac{3}{2}}$, $2d_{\frac{5}{2}}$, $3p_{\frac{1}{2}}$, $3p_{\frac{3}{2}}$, $3f_{\frac{5}{2}}$, $3f_{\frac{7}{2}}$.

Problem 3. (1) $(N+1)(N+2)$. (2) $\frac{\hbar^2}{2}N(N+3)$.
(3) $E_{Nlj} = \hbar\omega\left(\frac{\alpha_{Nlj}}{16} + \frac{3}{2}\right)$, where $\alpha_{Nlj} = 0(0s_{\frac{1}{2}})$, $10(1p_{\frac{3}{2}})$, $20(1p_{\frac{1}{2}})$,
$27(2d_{\frac{5}{2}})$, $37(2d_{\frac{3}{2}})$, $37(2s_{\frac{1}{2}})$, $39(3f_{\frac{7}{2}})$, $53(3f_{\frac{5}{2}})$, $53(3p_{\frac{3}{2}})$, $59(3f_{\frac{1}{2}})$.
(4) $l = N$, $j = N + \frac{1}{2}$.

Problem 4. $\varphi_n = \frac{1}{\sqrt{2\pi a}}\frac{1}{r}\sin\frac{n\pi r}{a}$, $E_n = \frac{1}{2M}\left(\frac{\hbar n\pi}{a}\right)^2$.

Problem 5. $r_{\max}^{(n_r=1,l=0)} = 5.2a_0$, $\langle 200|r|200\rangle = 6a_0$, $r_{\max}^{(n_r=0,l=1)} = 4a_0$,
$\langle 21m_l|r|21m_l\rangle = 5a_0$.

Problem 7. (1) $\frac{R}{\langle 100|r|100\rangle} = 1.5 \times 10^{-5}$ (H) ,
$\frac{R}{\langle 100|r|100\rangle} = 7.3 \times 10^{-3}$ (Pb) . (2) $\frac{R}{\langle 100|r|100\rangle} = 3.1 \times 10^{-3}$ (H) ,
$\frac{R}{\langle 100|r|100\rangle} = 1.5$ (Pb) .

Problem 8. $r^2 \to s$, $\varphi(r^2) \to s^{1/4}\phi(s)$, $l(l+1) \to \frac{1}{4}l(l+1) - \frac{3}{16}$,
$\frac{1}{4}E \to e^2/4\pi\epsilon_0$, $\frac{1}{8}M\omega^2 \to -E$.

Problem 9. $E_{s=0} = -\frac{3}{4}a\hbar^2$, $E_{s=1} = \frac{1}{4}a\hbar^2$.

Problem 10. (1) $E_{211\frac{1}{2}(-\frac{1}{2})} = 0$, $E_{210\frac{1}{2}\frac{1}{2}} = \mu_B B_z$.
(2) $E_{21\frac{1}{2}\frac{3}{2}\frac{1}{2}} = \hbar^2 v_{so}/2$, $E_{21\frac{1}{2}\frac{1}{2}\frac{1}{2}} = -\hbar^2 v_{so}$.
(3) $E_+ = \hbar^2 v_{so}(1+q)/2$, $E_- = -\hbar^2 v_{so}$.

Problem 11. $j_r = |A|^2 \hbar k / r^2 M$, flux$(d\Omega) = |A|^2 \hbar k d\Omega / M$.

Problem 12. (1) $V = V(\rho)$, $\rho \equiv \sqrt{x^2+y^2}$, $\phi \equiv \tan^{-1}(y/x)$.
(2) $\frac{1}{2M}\left(\hat{p}_x^2 + \hat{p}_y^2\right) = -\frac{\hbar^2}{2M}\left(\frac{\partial^2}{\partial \rho^2} + \frac{1}{\rho}\frac{\partial}{\partial \rho} + \frac{1}{\rho^2}\frac{\partial^2}{\partial \phi^2}\right)$, $E_{m_l} = E_{-m_l}$.
(3) $E_n = \hbar\omega(n+1)$, $n+1$, $n = 0, 1, 2, \ldots$.

Chapter 7

Problem 1. (1) $\frac{1}{2x_c\sqrt{2\pi}}\exp\left(-\frac{x^2}{2x_c^2}\right)\left(1+\frac{x^2}{x_c^2}\right)$, $x_c\sqrt{2}$, 0.10 .
(2) $\frac{1}{x_c\sqrt{2\pi}}\exp\left(-\frac{x^2}{2x_c^2}\right)$, x_c , 0.16 .
(3) $\frac{1}{x_c^3\sqrt{2\pi}}\exp\left(-\frac{x^2}{2x_c^2}\right)x^2$, $x_c\sqrt{3}$, 0.0021 .

Problem 2. (1) $\varphi_+ = \frac{1}{\sqrt{2}}\left[\varphi_{100}(1)\varphi_{21m_l}(2) + \varphi_{100}(2)\varphi_{21m_l}(1)\right]\chi_{s=0}$,
$\varphi_- = \frac{1}{\sqrt{2}}\left[\varphi_{100}(1)\varphi_{21m_l}(2) - \varphi_{100}(2)\varphi_{21m_l}(1)\right]\chi_{s=1,m_s}$. (2) $E_+ > E_-$.

Problem 3. $j = 0, 2, 4$.

Problem 4. (1) s. (2) a. (3) s. (4) a. (5) s.

Problem 5. j even.

Problem 6. (1) $3/2, 1/2$. (2) $1/2$.

Problem 8. $(\mathcal{N}, j, \Pi):$ $\left(1, \frac{1}{2}, +\right)$, $\left(3, \frac{3}{2}, -\right)$, $\left(5, \frac{3}{2}, -\right)$,
$\left(7, \frac{1}{2}, -\right)$, $\left(9, \frac{5}{2}, +\right)$, $\left(13, \frac{5}{2}, +\right)$.

Problem 9. (1) $3.8/-0.26/4.8$ (μ_p). (2) $-1.9/0.64/-1.9$ (μ_p).

Problem 10. $n(\epsilon) = \frac{M\epsilon}{\pi\hbar^2}$, $C_V = 2n_F k_B \frac{T}{T_F}$.

Problem 11. (1) 5.9×10^3 Å. (2) Red.

Problem 12. $q_{aa} + q_{bb}$, $\quad -q_{ac}$, $\quad q_{bc}$.

Chapter 8

Problem 1. (1) Equation (8.10).

(2) $c_{p \neq n}^{(2)} = \dfrac{1}{E_n^{(0)} - E_p^{(0)}} \left[\sum_{q \neq n} c_q^{(1)} \langle \varphi_p^{(0)} | V | \varphi_q^{(0)} \rangle - E_n^{(1)} c_p^{(1)} \right]$,

$c_n^{(2)} = -\dfrac{1}{2} \sum_p |c_p^{(1)}|^2$.

Problem 2. (1) $E_1^{(1)} = E_2^{(1)} = 0$, $\quad E_3^{(1)} = 2c$.

(2) $E_1^{(2)} = -E_2^{(2)} = \dfrac{|c|^2}{3}$, $\quad E_3^{(2)} = 0$.

(3) $\varphi_1^{(1)} = \dfrac{c}{3} \varphi_2^{(0)}$, $\quad \varphi_2^{(1)} = -\dfrac{c}{3} \varphi_1^{(0)}$, $\quad \varphi_3^{(1)} = 0$.

(4) $\varphi_1^{(2)} = -\dfrac{|c|^2}{18} \varphi_1^{(0)}$, $\quad \varphi_2^{(2)} = -\dfrac{|c|^2}{18} \varphi_2^{(0)}$, $\quad \varphi_3^{(2)} = 0$.

(5) $E_\pm = \dfrac{7}{2} \pm \dfrac{3}{2} \sqrt{1 + \dfrac{4|c|^2}{9}} \approx \dfrac{7}{2} \pm \dfrac{|c|^2}{3}$, $\quad E_3 = -1 + 2c$.

Problem 3. (1) $E_0^{(1)} = 0$, $\quad E_0^{(2)} = -\dfrac{k^2}{2M\omega^2}$.

(2) $E_0^{(1)} = \dfrac{bx_c^2}{4}$, $\quad E_0^{(2)} = -\dfrac{b^2 x_c^2}{16 M \omega^2}$.

Problem 4. $\Delta E = \dfrac{3 \hbar \omega}{4} \left(\dfrac{x_c}{R_0} \right)^4 l(l+1) - \dfrac{\hbar \omega}{2} \left(\dfrac{x_c}{R_0} \right)^6 l^2(l+1)^2$.

Problem 5. (1) $E_0^{(1)} = -\dfrac{3}{32M} \left(\dfrac{\hbar \omega}{c} \right)^2$. (2) 10^{-8}.

Problem 6. $\Psi_n = \left[1 - \dfrac{1}{2} \sum_{p \neq n} \dfrac{|\langle \varphi_p^{(0)} | V | \varphi_n^{(0)} \rangle|^2}{(E_n^{(0)} - E_p^{(0)})^2} \right] \varphi_n^{(0)}$

$+ \sum_{p \neq n} \dfrac{\langle \varphi_p^{(0)} | V | \varphi_n^{(0)} \rangle}{E_n^{(0)} + \langle \varphi_n^{(0)} | V | \varphi_n^{(0)} \rangle - E_p^{(0)}} \varphi_p^{(0)} + \sum_{p,q(\neq n)} \dfrac{\langle \varphi_p^{(0)} | V | \varphi_q^{(0)} \rangle \langle \varphi_q^{(0)} | V | \varphi_n^{(0)} \rangle}{(E_n^{(0)} - E_p^{(0)})(E_n^{(0)} - E_q^{(0)})} \varphi_p^{(0)}$.

Problem 8.

(1) $\langle H \rangle = \dfrac{\hbar \omega}{4} \left[\dfrac{M}{M^*} + \dfrac{M^*}{M} - \dfrac{3}{8} \dfrac{\hbar \omega}{Mc^2} \left(\dfrac{M^*}{M} \right)^2 + \dfrac{7}{16} \left(\dfrac{\hbar \omega}{Mc^2} \right)^2 \left(\dfrac{M^*}{M} \right)^3 + \cdots \right]$.

(2) $1 = \left(\dfrac{M^*}{M} \right)^2 - \dfrac{3}{4} \dfrac{\hbar \omega}{Mc^2} \left(\dfrac{M^*}{M} \right)^3 + \dfrac{21}{16} \left(\dfrac{\hbar \omega}{Mc^2} \right)^2 \left(\dfrac{M^*}{M} \right)^4 + \cdots$.

(3) $\dfrac{M^*}{M} = 1 + \dfrac{3}{8}\dfrac{\hbar\omega}{Mc^2} - \dfrac{39}{64}\left(\dfrac{\hbar\omega}{Mc^2}\right)^2 + \cdots$.

(4) $\langle H \rangle = \dfrac{\hbar\omega}{2}\left[1 - \dfrac{3}{16}\dfrac{\hbar\omega}{Mc^2} + \dfrac{19}{128}\left(\dfrac{\hbar\omega}{Mc^2}\right)^2 + \cdots\right]$.

Problem 9. $\left\langle 1s2p\pm \left| \dfrac{e^2}{4\pi\epsilon_0 r} \right| 1s2p\pm \right\rangle = -(0.98 \pm 0.08)E_{\mathrm{H}}$.

Problem 10.

	$\langle H \rangle_Z$	Z^*	$\langle H \rangle_{Z^*}$	exp
He	5.50	1.69	5.69	5.81
Li$^+$	14.25	2.69	14.44	14.49
Be^{++}	27.00	3.69	27.19	27.21

Problem 11. $E_\pm(m=0) = E^{(0)}_{n=2} \pm 3eE_z a_0$, $\quad E(m=\pm 1) = E^{(0)}_{n=2}$.

Problem 12. (1) $j(2j+1) \times j(2j+1)$. (2) $\left(j+\dfrac{1}{2}\right) \times \left(j+\dfrac{1}{2}\right)$.

(3) $E_a = -g\left(j+\dfrac{1}{2}\right)$, $\quad E_b = 0$.

Chapter 9

Problem 1. $0.50 - 0.40\sin(3\pi^2 \hbar t/2Ma^2)$.

Problem 2. $\dfrac{\mathrm{d}\langle\Psi|p|\Psi\rangle}{\mathrm{d}t} = -\left\langle \Psi \left| \dfrac{\mathrm{d}V}{\mathrm{d}x} \right| \Psi \right\rangle$.

Problem 3. $c_{y\uparrow} = \dfrac{1-\mathrm{i}}{\sqrt{2}}\cos\left(\dfrac{1}{2}\omega_\mathrm{L} t + \dfrac{\pi}{4}\right)$.

Problem 4. $0.36, 0.50, 0.13$.

Problem 5. $P^{(1)}_{\uparrow\to\downarrow} = \dfrac{\omega'^2}{(\omega-\omega_\mathrm{L})^2}\sin^2\left[\dfrac{1}{2}t(\omega-\omega_\mathrm{L})\right]$.

Problem 6.

(1) $c_{0\to 1} = -\dfrac{\mathrm{i}vV_0}{\hbar x_\mathrm{c}}\exp\left(\dfrac{\hbar\omega}{4Mv^2}\right)\displaystyle\int_{t_1}^{t_2} t\exp\left[-\dfrac{v^2}{x_\mathrm{c}^2}\left(t - \mathrm{i}\dfrac{\hbar\omega}{2Mv^2}\right)\right]\mathrm{d}t$.

(2) $|c_{0\to 1}|^2 = \dfrac{V_0\omega^2}{2Mv^2}\exp\left(\dfrac{\hbar\omega}{2Mv^2}\right)$.

Problem 7.

(1) $\Psi(t) = \cos\theta_0 \exp\left[-\dfrac{\mathrm{i}V_0\sin(\omega t)}{4\hbar\omega}\right]\chi^1_0 + \sin\theta_0 \exp\left[\dfrac{\mathrm{i}3V_0\sin(\omega t)}{4\hbar\omega}\right]\chi^0_0$.

(2) $\Psi(t) = \exp\left[-\frac{iV_0 \sin(\omega t)}{4\hbar\omega}\right]\varphi_{B_0}$.

Problem 8. $P_{0\to 1} = 2\left(\dfrac{Kx_c}{\hbar\omega}\right)^2 \sin^2\dfrac{\omega t}{2}$.

Problem 9.
(1) $c_k^{(2)} = \dfrac{1}{\hbar^2}\sum_j \langle k|V|j\rangle\langle j|V|i\rangle \left[\dfrac{1}{\omega_{ki}\omega_{kj}} + \dfrac{\exp(i\omega_{ki}t)}{\omega_{ki}\omega_{ji}} + \dfrac{\exp(i\omega_{kj}t)}{\omega_{kj}\omega_{ij}}\right]$.

(2) $P_{0\to 2} = 2\left(\dfrac{Kx_c}{\hbar\omega}\right)^4 \sin^4\dfrac{\omega t}{2}$.

Problem 10. (1) 10^5 . (2) 10^2 .

Problem 11. $\left|\dfrac{\langle 210|r|100\rangle}{\langle 310|r|100\rangle}\right|^2 = 6.3$, $\dfrac{P(100\to 210)}{P(100\to 310)} = 3.8$.

Problem 12. (1) $\dfrac{P(310\to 200)}{P(310\to 100)} = 0.13$. (2) 1.1×10^{-8} s. (3) 4×10^{-7} eV.

Chapter 10

Problem 1. Real amplitudes.

Problem 2. 42%.

Problem 3. $\hat{S}_z(\varphi_{B_1})$, $\hat{S}_x(\varphi_{B_2})$, $\hat{S}_y(\varphi_{B_3})$.

Problem 4. $\mathcal{U}_H$.

13.2 Physical Units and Constants

The table on the following page shows the equivalence between physical units and the value of constants used in the text.

Table 13.1. Equivalence between physical units and the value of constants used in the text [101]

Quantity	Symbol	Units (m.k.s.)	Atomic scale	Nuclear scale
		1 m	10^{10} Å	10^{15} F
		1 J	0.625×10^{19} eV	0.625×10^{13} MeV
		1 kg	0.56×10^{36} eVc^{-2}	0.56×10^{30} MeVc^{-2}
Bohr magneton	μ_B	0.93×10^{-23} JT^{-1}	0.58×10^{-4} eVT^{-1}	0.58×10^{-10} MeVT^{-1}
Bohr radius	a_0	0.53×10^{-10} m	0.53 Å	0.53×10^5 F
Boltzmann constant	k_B	1.38×10^{-23} JK^{-1}	0.86×10^{-4} eVK^{-1}	0.86×10^{-10} MeVK^{-1}
Constant in Coulomb law	$e^2/4\pi\epsilon_0$	2.34×10^{-28} Jm	14.4 eVÅ	1.44 MeVF
Electron charge	e	-1.60×10^{-19} C		
Electron mass	M, M_e	0.91×10^{-30} kg	0.51×10^6 eVc^{-2}	0.51 MeVc^{-2}
Fine structure constant	$\alpha = e^2/4\pi\epsilon_0\hbar c$	1/137		
Hydrogen atom ground state	E_H	-2.18×10^{-18} J	-13.6 eV	-1.36×10^{-5} MeV
Deuteron nucleus ground state	E_D	-3.57×10^{-13} J	-2.23×10^6 eV	-2.23 MeV
Nuclear magneton	μ_p	0.51×10^{-26} JT^{-1}	0.32×10^{-7} eVT^{-1}	0.32×10^{-13} MeVT^{-1}
Planck constant/2π	$\hbar$	1.05×10^{-34} Js	0.66×10^{-15} eVs	0.66×10^{-21} MeVs
Proton mass	M_p	1.67×10^{-27} kg	0.94×10^9 eVc^{-2}	0.94×10^3 MeVc^{-2}
Rydberg constant	R_H	1.10×10^7 m^{-1}	1.10×10^{-3} Å^{-1}	1.10×10^{-8} F^{-1}
Speed of light in vacuum	c	3.00×10^8 ms^{-1}	3.00×10^{18} Ås^{-1}	3.00×10^{23} Fs^{-1}

References

1. B.C. Olschak: *Buthan. Land of Hidden Treasures*. Photography by U. and A. Gansser (Stein & Day, New York 1971)
2. M. Planck: Verh. Deutsch. Phys. Ges. **2**, 207, 237 (1900)
3. A. Einstein: Ann. der Phys. **17**, 132 (1905)
4. A.H. Compton: Phys. Rev. **21**, 483 (1923)
5. C.L. Davisson and L.H. Germer: Nature **119**, 528 (1927); G.P. Thomson: Proc. Roy. Soc. A **117**, 600 (1928)
6. H. Geiger and E. Mardsen: Proc. Roy. Soc. A **82**, 495 (1909); E. Rutherford: Phil. Mag. **21**, 669 (1911)
7. J. Balmer: Verh. Naturf. Ges. Basel **7**, 548, 750 (1885); Ann. der Phys. und Chem. **25**, 80 (1885)
8. N. Bohr: Phil. Mag. **25**, 10 (1913); **26**, 1 (1913); Nature **92**, 231 (1913)
9. W. Heisenberg: Zeitschr. Phys. **33**, 879 (1925)
10. M. Born, W. Heisenberg and P. Jordan: Zeitschr. Phys. **35**, 557 (1926)
11. P.A.M. Dirac: Proc. Roy. Soc. A **109**, 642 (1925)
12. E. Schrödinger: Ann. der Phys. **79**, 361, 489 (1926); **80**, 437 (1926); **81**, 109 (1926)
13. P.A.M. Dirac: Proc. Roy. Soc. A **117**, 610 (1928); A **118**, 351 (1928)
14. R.P. Feynman: Rev. Mod. Phys. **20**, 367 (1948); R.P. Feynman and A.R. Hibbs: *Quantum Mechanics and Path Integrals* (McGraw-Hill, New York, St. Louis, San Francisco, London 1965)
15. J. Schwinger: *Quantum Mechanics. Symbolism of Atomic Measurements*, ed. by B.G Englert (Springer-Verlag, Berlin, Heidelberg, New York 2001) Chap. 1
16. P.A.M. Dirac: *The Principles of Quantum Mechanics*, 3rd edn. (Oxford University Press, Amen House, London 1930) Chaps. 1–5
17. R.F. Feynman, R.B. Leighton and M. Sands: *The Feynman Lectures on Physics. Quantum Mechanics* [Addison-Wesley, Reading (Massachusetts), London, New York, Dallas, Atlanta, Barrington (Illinois) 1965] Chap. 5
18. J.J. Sakurai: *Modern Quantum Mechanics*, ed. by San Fu [Addison-Wesley, Reading (Massachusetts), London, New York, Dallas, Atlanta, Barrington (Illinois) 1994] Chap. 1
19. A. Einstein, B. Podolsky and N. Nathan: Phys. Rev. **47**, 777 (1935)
20. Å. Bohr and O. Ulfbeck: Rev. Mod. Phys. **67**, 1 (1995)
21. D.F. Styer: Am. J. Phys. **64**, 31 (1996)
22. A.P. French and E.F. Taylor: *An Introduction to Quantum Mechanics* (W.W. Norton & Co., New York 1978) Chap. 6
23. O. Nairz, M. Arndt and A. Zeilinger: Am. J. Phys. **71**, 319 (2003)
24. J. Roederer: Entropy **5**, 3–33 (2003); *Information and its Role in Nature* (Springer-Verlag, Berlin, Heidelberg, New York, to appear in 2005)

25. N.D. Mermin: Am. J. Phys. **71**, 23 (2003)
26. B. Cougnet, J. Roederer and P. Waloshek: Z. fur Naturforshung A **7**, 201 (1952)
27. M. Born: Zeitschr. Phys. **37**, 863 (1926); **38**, 499 (1926)
28. L. de Broglie: C. R. Acad. Sci. Paris **177**, 507, 548 (1923)
29. F. Bloch: Zeitschr. Phys. **52**, 555 (1928)
30. O. Stern and W. Gerlach: Zeitschr. Phys. **8**, 110 (1921); **9**, 349 (1922)
31. G.E. Uhlenbeck and S.A. Goudsmit: Nature **113**, 953 (1925); **117**, 264 (1926)
32. W. Pauli: Zeitschr. Phys. **43**, 601 (1927)
33. D.A. McQuarrie: *Quantum Chemistry* (University Science Books, Herndon, Virginia 1983) Fig. 6-12
34. A.K. Grant and J.L. Rosner: Am. J. Phys. **62**, 310 (1994)
35. F. Wilczek: Scientific American **264**, Vol. 5, 24 (1991); *Fractional Statistics and Anyon Superconductivity* (World Scientific, Singapore, New Jersey, London, Hong Kong 1990)
36. R.H. Romer: Am. J. Phys. **70**, 791 (2002)
37. W. Pauli: Zeitschr. Phys. **31**, 625 (1925)
38. Å. Bohr and B. Mottelson: *Nuclear Structure* (W.A. Benjamin, New York, Amsterdam 1969) Vol. I, Chaps. 2–4
39. S.N. Bose: Zeitschr. Phys. **26**, 178 (1924)
40. P.A.M. Dirac: Proc. Roy. Soc. A **112**, 661 (1926)
41. E. Fermi: Zeitschr. Phys. **36**, 902 (1926)
42. D. Kleppner: Physics Today **49**, 11 (1996); F. Dalfovo, S. Giorgini, L.P. Pitaevckii and S. Stringari: Rev. Mod. Phys. **71**, 463 (1999)
43. F. London: Jour. Phys. Chem. **43**, 49 (1939)
44. A. Einstein: Sitz. Ber. Preuss. Ak. Wiss. 61 (1924); 3 (1925)
45. M.H. Anderson, J.R. Enscher, M.R. Matthews, C.E. Wieman and E.A. Cornell: Science **269**, 198 (1995); J.R. Enscher, D.S. Jin, M.R. Matthews, C.E. Wieman and E.A. Cornell: Phys. Rev. Lett. **77**, 4984 (1996)
46. K. von Klitzing, G. Dorda and M. Pepper: Phys. Rev. Lett. **45**, 494 (1980)
47. D.C. Tsui, H.L. Störmer and A.C. Gossard: Phys. Rev. Lett. **48**, 1559 (1982); Phys. Rev. B **25**, 1405 (1982)
48. R.E. Prange: Introduction to *The Quantum Hall Effect*, ed. by R.E. Prange and S.M. Girvin (Springer-Verlag, New York, Berlin, Heidelberg 1987) Fig. 1.2
49. B.L. Halperin: Scientific American **254**, Vol. 4, 52 (1986)
50. J.D. Jackson: *Classical Electrodynamics* (John Wiley & Sons, New York, Chichester, Brisbane, Toronto 1975) p. 574
51. R.B. Laughlin: Phys. Rev. Lett. **50**, 1395 (1983)
52. R.P. Feynman: Phys. Rev. **76**, 769 (1949)
53. C. Itzykson and J.B. Zuber: *Quantum Field Theory* (McGraw-Hill, New York, St. Louis, San Francisco, London 1980)
54. J. Bardeen, L.N. Cooper and J.R. Schrieffer: Phys. Rev. **106**, 162 (1957)
55. A. Einstein: Verh. Deutsch. Phys. Ges. **18**, 318 (1916); Mitt. Phys. Ges. Zürich **16**, 47 (1916); Phys. Zeitschr. **18**, 121 (1917)
56. B.H. Brandow: Rev. Mod. Phys. **39**, 771 (1967); E.M. Krenciglowa and T.T.S. Kuo: Nucl. Phys. A **240**, 195 (1975)
57. C. Bloch and J. Horowitz: Nucl. Phys. **8**, 91 (1958)
58. C. Becchi, A. Rouet and R. Stora: Phys. Lett. B **52**, 344 (1974); Ann. Phys. **98**, 287 (1976); I.V. Tyutin: Lebedev preprint FIAN, **39** (1975)

59. M. Henneaux and C. Teitelboim: *Quantization of Gauge Systems* (Princeton University Press, Princeton, New Jersey 1992)
60. D.R. Bes and J. Kurchan: *The Treatment of Collective Coordinates in Many-Body Systems. An Application of the BRST Invariance* (World Scientific Lecture Notes in Physics, Vol. 34, Singapore, New Jersey, London, Hong Kong 1990)
61. D.R. Bes and O. Civitarese: Am. J. Phys. **70**, 548 (2002)
62. B.M. Terhal, M.M. Wolf and A.C. Doherty: Physics Today **56**, 46 (2003)
63. C.H. Bennet and G. Brassard: Proc. IEEE Int. Conf. on Computers, Systems and Signal Processing (IEEE Press, Los Alamos, California 1984) p. 175
64. D. Gottesman and H.-K. Lo: Physics Today **53**, 22 (2000)
65. W. Wootters and W. Zurek: Nature **299**, 802 (1982)
66. C.H. Bennett, G. Brassard, C. Crépeau, R. Jozsa, A. Peres and W. Wooters: Phys. Rev. Lett. **70**, 1895 (1993)
67. D. Bouwmeester, J.W. Pan, K. Mattle, M. Eibl, H. Weinfurter and A. Zeilinger: Nature **390**, 575 (1997)
68. P.W. Schor: Proc. 34th Annual Symp. Found. Comp. Scien. (FOCS), ed. by S. Goldwasser (IEEE Press, Los Alamitos, California 1994) p. 124
69. I.L. Chuang, L.M.K. Vandersypen, X.L. Zhou, D.W. Leung and S. Lloyd: Nature **393**, 143 (1998)
70. M.A. Nielsen and I.L. Chuang: *Quantum Computation and Quantum Information* (Cambridge University Press, Cambridge 2001) p. 532
71. P.A.M. Dirac: Hungarian Ac. of Sc. Rep. KFK-62 (1977)
72. N. Bohr: Nature **121** (suppl.), 580 (1928)
73. N. Bohr: Nature **136**, 65 (1935); Phys. Rev. **48**, 696 (1935)
74. H. Everett III: Rev. Mod. Phys. **29**, 315 (1957)
75. J.L. Borges: The Garden of Forking Paths. In: *Labyrinth. Selected Stories and Other Writings* (New Directions Publishing, New York 1961) p. 26. First Spanish edition: *El Jardín de los Senderos que se Bifurcan* (Sur, Buenos Aires 1941); A.G. Rojo: *El Jardín de los Mundos que se Ramifican: Borges y la Mecánica Cuántica*, in *Borges en 10 Miradas* (Fundación El Libro, Buenos Aires 1999) p. 185
76. J.P. Paz and W.H. Zurek: Environment-Induced Decoherence and the Transition from Classical to Quantum. In: *Coherent Atomic Matter Waves*, Les Houches Session LXXXII, ed. by R. Kaiser, C. Westbrook and F. Davids (Springer-Verlag, Berlin, Heidelberg, New York 2001) p. 533
77. J. von Neumann: *Matematische Grundlagen der Quantenmechanik* (Springer, Berlin 1932)
78. M. Tegmark and J.A. Wheeler: Scientific American **284**, Vol. 2, 54 (2001)
79. E. Hobsbawn: *The Age of Extremes. A History of the World, 1914–1991* (Vintage Books, New York 1996) p. 7
80. H. Kragh: *Quantum Generations. A History of Physics in the Twentieth Century* (Princeton University Press, Princeton, New Jersey 1999)
81. A. Pais: *Inward Bound* (Oxford University Press, Oxford, New York, Toronto 1986); *Niels Bohr's Times. In Physics, Philosophy and Politics* (Clarendon Press, Oxford 1991)
82. P. Robertson: *The Early Years. The Niels Bohr Institute 1921–1930* (Akademisk Forlag. Universitetsforlaget i København, Denmark 1979)

83. O. Spengler: *Der Untergang des Abendlandes. Umrisse einer Morphologie der Weltgeschichte*, Vol. 1: *Gestalt und Wirklichkeit* (Munich, 1918). First English translation: *The Decline of the West*, Vol. 1: *Form and Actuality* (Knof, New York 1926)
84. E. Rutherford: Phil. Mag. **49**, 1 (1900)
85. H.G.J. Moseley: Nature **92**, 554 (1913); Phil. Mag. **26**, 1024 (1913); **27**, 703 (1914)
86. J. Frank and H. Hertz: Verh. Deutsch. Phys. Ges. **16**, 457 (1914)
87. A. Sommerfeld: Sitz. Ber. Bayer. Akad. Wiss. 459 (1915)
88. L.H. Thomas: Nature **117**, 514 (1926); Phil. Mag. **3**, 1 (1927)
89. M. Born and P. Jordan: Zeitschr. Phys. **34**, 858 (1925)
90. W. Pauli: Zeitschr. Phys. **36**, 336 (1926)
91. P.A.M. Dirac: Proc. Roy. Soc. A **110**, 561 (1926)
92. W. Heisenberg: Zeitschr. Phys. **38**, 499 (1926)
93. W. Heisenberg: Zeitschr. Phys. **43**, 172 (1927)
94. N. Bohr: *The Philosophical Writings of Niels Bohr* (Ox Bow Press, Woodbridge, Connecticut 1987)
95. N. Bohr: Discussions with Einstein on Epistemological Problems in Atomic Physics. In *Albert Einstein. Philosopher-Scientist*, ed. by P.A Schilpp (Open Court Publishing, Peru, Illinois 2000) p. 199
96. A. Einstein: Reply to Criticisms. In: *Albert Einstein. Philosopher-Scientist*, ed. by P.A Schilpp (Open Court Publishing, Peru, Illinois 2000) p. 663
97. D. Bohm: Phys. Rev. **85**, 166, 180 (1952)
98. J.S. Bell: Physics **1**, 195 (1964)
99. A. Aspect, P. Grangier and G. Roger: Phys. Rev. Lett. **47**, 460 (1981); **49**, 91 (1982); A. Aspect, J. Dalibard and G. Roger: Phys. Rev. Lett. **49**, 1804 (1982)
100. G. Weihs, T. Jennewein, C. Simon, H. Weinfurter and A. Zeilinger: Phys. Rev. Lett. **81**, 5039 (1998)
101. The NIST Reference on Constants, Units, and Uncertainty: Appendix 3. http:/physics.nist.gov/cuu/Units/units.html

Index

absorption processes   148
adjoint vector   25
Ag atom   68
allowed transitions   150
alpha particle   97
angular momentum
   addition   70, 75
   commutation relations   63, 64
   matrix treatment   63–65, 72, 73
   orbital   64–67, 73, 74
anomalous Zeeman effect   93
anthropocentric foundation of quantum mechanics   168, 169
anticommutation relation   107
antisymmetric states   95, 96
anyons   96, 116
apparatus   12, 15, 170
average   22

Balmer series   175, 176
band structure of crystals   57, 105, 130
Be atom   135
Bell
   inequalities   184
   states   156, 162
Bell, J.   184
Bessel functions   84, 90, 91
black-body radiation   1, 175
Bloch theorem   57
Bloch–Horowitz diagonalization procedure   128
Bohm, D.   184
Bohr
   Festspiele   180
   frequency   142
   magneton   67, 68, 70, 193
   model   1, 176, 177
   radius   80, 87, 193

Bohr, N.   X, 1, 2, 69, 168, 169, 174, 176–178, 180–183
Boltzmann constant   104, 193
Boltzmann, L.   1
Born, M.   1, 174, 177–180
Born–Oppenheimer approximation   123
Bose, S.   177
Bose–Einstein
   condensation   98, 109–111, 177
   distribution   108, 177
boson   96, 97, 116
   occupation number   97, 106
boundary conditions   44, 45, 49, 51, 83, 84, 127
bra   9
Brillouin–Wigner perturbation theory   121, 128, 134
BRST   133

center of mass   125, 126
central potentials   79, 80
classical computation   154, 157
classical electromagnetism   6, 67, 145, 146
classical physics   5, 6, 63, 154
Clebsch–Gordan coefficients   71, 75
closed shell   100, 102, 103, 105, 117
closure   14, 21, 26, 34, 37, 60
collective
   coordinate   131, 133
   subspace   133
column vector   25
commutation relation   8, 10, 17
complementarity   2, 181
complete set of states   7, 21, 40, 44, 66, 69

completeness of quantum mechanics 183
Compton effect 1, 175
conduction band 105
conductor 105
confluent hypergeometric functions 89
constant-in-time perturbation 143
continuity conditions 49, 53–55, 59, 85
continuity equation 42
control qubit 159
controlling Hamiltonian 158, 164
Cooper pairs 136
Copenhagen interpretation 11, 168, 179
correspondence principle 45, 176
Coulomb
  potential 80–82, 87, 88, 91, 100, 101, 122
  repulsion 99, 116, 122, 124
covalent binding 123
creation and annihilation operators 32, 33, 98, 106, 107, 118, 146
cross-section
  differential 85
  total 85
cryptographic key 154
cryptography IX, 154

de Broglie, L. 2, 179
decay law 150, 175
decoherence 154, 162, 167, 169–171
degenerate states 48, 50, 57, 80, 81, 91–93, 99, 113, 120, 124, 128, 129, 135
density of states 104, 144, 147
detector 12, 15
determinism 1, 179
deterministic evolution of the state vector 167
deuteron 117
diagonal matrix element 9
differential formulation 2, 40, 179
dilatation 8
dimensionless coordinates 43, 46, 82, 87
Dirac equation 2, 82, 179
Dirac, P.A.M. IX, 1, 9, 142, 176–180
distance of closest approach 86

Ehrenfest theorem 151
Ehrenfest, P. 177
eigenfunction 44, 49, 50, 55, 61
eigenstate 8, 11, 12, 16, 18–21
eigenvalue 8, 9, 11, 16–18, 20, 27, 29, 44, 47, 49, 50, 56, 57
eigenvalue equation 8, 11–13, 40, 55, 60
eigenvector 8, 9, 11, 16, 17, 21, 23
Einstein, A. 1, 5, 109, 149, 175–177, 181–183
electron
  charge 193
  diffraction 1
  gas 49, 103
  mass 193
electronic shell structure 99, 100
emission processes 148
entangled
  photons 157, 184
  states 96, 99, 153, 154, 163, 182, 184
entanglement IX, 153–155, 158
environment 158, 170
EPR 5, 154, 168, 182–184
Euclid 161
Euclidean space 7
Everett III, H. 168
evolution operator 137, 138
exclusion principle 97, 107, 177, 179
expectation value 16, 18, 23, 29, 32, 37, 43, 60, 61, 120, 122, 135

factorization 79, 95, 158
Faraday, M. 1
Fermi
  energy 103, 109, 114
  golden rule 144, 148
  momentum 103
  temperature 104
Fermi, E. 177
Fermi–Dirac distribution 105, 108, 177
fermion 96, 97, 117, 118
  occupation number 97, 106
Feynman perturbation theory 121
Feynman, R. IX, 2, 15, 97
filter 12
final state 139, 142–144, 149, 157
fine structure constant 147, 193

finite square well 39, 55
folded diagrams 128
forbidden transitions 150
Fourier expansion 7
Fourier transformation 160, 161, 165
   classical 160
fourth quantum number 68, 177
fractional quantum Hall effect 115
Frank, J. 176
free particle 7, 19, 47
   time-dependent state 5, 138
fullerenes 15

Galileo, G. 1
Gamow, G. 178
Geiger, H. 175
Gerlach, W. 12, 68, 177
Gossard, A. 112
Goudsmit, S. 68, 177
gyromagnetic ratio 67, 70

H atom 1, 80, 81, 83, 87, 88, 92, 121, 122, 135
H ion 29, 123–125
Hall effects 112, 115
Hamiltonian 11, 23, 31, 34, 40, 43, 44, 79, 80, 138
Hankel functions 84, 90
hard sphere scattering 86
harmonic oscillator 101
   1D 89, 116, 127, 134, 135
     matrix treatment 31, 33–35, 178
     position treatment 39, 43–45, 47
   2D 93, 112
   3D 80, 81, 88, 89, 91, 92, 101, 102, 109, 117, 135
harmonic oscillators, 1D, matrix treatment 34
He atom 98, 116, 122, 135, 177, 179
He nuclei 97
Heisenberg realization of quantum mechanics 25, 177, 181
Heisenberg, W. 1, 2, 17–19, 174, 177–181
Helmholtz, H. 1
Hermite polynomials 44
Hermitian
   conjugate 18, 20, 22
   operator 9, 11, 17, 18, 20, 21, 44

Hertz, H. 176
hidden variables 183, 184
Hilbert space 7–9, 12, 178
history of quantum mechanics 173–184
Hobsbawn, E. 173
hybridization 37
hyperfine interaction 83

identical particles 95
indeterminacy 2
induced emission 149
infinite potential well 39, 49
initial state 12, 15, 139–142, 144, 149, 157
insulator 105
integer quantum Hall effect 112
interaction 15
   of light with particles 147, 148
interference 14, 158
intrinsic coordinate 131, 133

j-shell 117, 136
Jordan, P. 1, 177, 178, 180

ket 9
Klein, O. 176, 178
Kramers, H. 176–178
Kronig, R. 177

Laguerre polynomial 87
Lamb shift 121
Landau levels 113, 115
Landau, L. 178
Laplacian 79
Larmor frequency 139
Laughlin states 116
Laughlin, R. 116
Legendre
   functions 74
   polynomials 74, 130
Lenard, P. 174
Levi–Civita tensor 64
Li atom 135
linear independence 7
locality of quantum mechanics 184
Lorentz force 111
Lorentz, H. 177

magnetic
  moment   67
  resonance   139, 141
  trap   109
Malus law   13
many-worlds interpretation   168
Mardsen, E.   175
matrix
  2×2 28
  as operator   26
  diagonal   27
  diagonalization   27, 29, 128
  eigenvalue equation   26, 27
  element   9
  elements of $1/r_{12}$ 130
  formulation   1, 2, 25, 178, 179
  multiplication   26
  unitary   28
matrix treatment   34
Maxwell, J.   1, 176
Maxwell–Boltzmann distribution   108, 110, 111
mean lifetime   150
mean square deviation   22
mean value   16
measurement   6, 9, 11, 12, 14–16, 154, 167, 168, 181
Mendeleev chart   100
Mermin, N.D.   17
meson   97, 152
Millikan, R.   175
molecule   123–126, 128
  intrinsic motion   123, 125
  rotational motion   126, 134
  vibrational motion   127, 128, 134
momentum
  distribution   110
  eigenfunction   47, 50, 103
  eigenvalue   47, 50, 103
  operator   40
Moseley, H.   176
muon   92, 97

n-qubit system   160
Na electron gas   104
Neumann functions   84, 90
neutrino   97
neutron   97, 101–103, 117
Newton's second law   6, 151

Newton, I.   1, 176
Niels Bohr Institutet   X, 176
no-cloning theorem   156, 163
no-crossing rule   29
non-commutative algebra   8
non-diagonal matrix elements   149
Nordita   X
norm   7, 8, 20, 22
normalization   27, 39, 44, 120
Notgemeinschaft der Deutschen Wissenschaft   174
nuclear
  magneton   70, 193
  shell structure   101, 102
nucleon   101, 103

observable   11, 15
occupation number   97, 106
old quantum theory   174
one-qubit
  gate   159, 164
  system   158
one-step potential   39, 51
operator   11, 16
  in differential form   40
  in matrix form   26
optical theorem   86
orthogonal vectors   7
orthonormalization   7, 21

parahelion precession   176
parity   35, 49, 66, 81
Pauli matrices   69
Pauli principle   96, 100, 104, 105
Pauli, W.   97, 177–180
Pb atom   92
periodic boundary conditions   50, 59
periodic potential   39, 57, 105, 129
perturbation theory   119–121, 133
  Feynman   121
  time-dependent   141–144
phase shift   84
phase, overall   10
photoelectric effect   1, 175
photon   12, 97, 144, 148, 152, 175, 182, 184
physical operators and states   132
physical quantity   15
physical reality   5, 182

pion   20
Planck constant   10, 175, 193
Planck radiation law   146, 149, 175
Planck, M.   174–177
plane wave   5, 47, 50, 103, 138, 145
Podolsky, B.   5, 182
pointer states   169, 170
Poisson bracket   178
polarized photons   12, 145, 148, 155
positronium   92
probabilistic interpretation   11, 13, 16, 41, 179
probability   17
  current   41, 42
  density   22, 41, 42, 45, 46, 67, 87, 109
projection operator   21
proton   23, 97, 101–103, 117, 193
publications per country   180

quantization of the electromagnetic field   146, 147
quantum computation   154, 157, 161, 162
quantum electrodynamics   144
quantum gates   158, 170
quantum Hall effects   96, 111–116
quantum information   153
quantum network   158
quantum register   160
qubit   70, 155–157, 159, 164

Raleigh–Schrödinger perturbation theory   121, 128
Rb atom   110
reduced mass   87, 125
reduction of the state vector   16, 137, 168
reflection coefficient   52, 54, 57
relativistic correction   134
Rn^{220} nucleus   174
root mean square   22
root mean square deviation   34
Rosen, N.   5, 182
rotations   8, 65, 126
running waves   49, 103
Rutherford, E.   174, 175
Rydberg constant   176, 193

scalar product   7, 25, 39

scattering   39, 83–86
  theory   84
Schrödinger equation   2
  time-dependent   41, 138, 141, 179, 184
  time-independent   39, 40, 42, 43, 47, 179, 184
Schrödinger realization of quantum mechanics   39, 179, 181
Schrödinger, E.   2, 179, 180
Schwinger, J.   IX, 2, 6
selection rules   149, 150
semiconductor   105
separation of variables   79
shell   100
Slater determinant   97, 106, 107, 115, 117
social context   173
Solvay conferences   181, 182
Sommerfeld, A.   176, 180
specific heat of metals   104
speed of light   193
spherical coordinates   65, 79
spherical harmonics   66, 74, 79
spherical wave   84
spin   28, 30, 63, 69, 70, 75, 76, 177, 179
  time dependence   138, 139
spin flip   140
spin–orbit interaction   83, 101
spin–statistics theorem   97
spontaneous emission   149
spread in energy   144
square barrier   39, 54
Störmer, H.   112
standard deviation   16, 18
Stark effect   135
Stark, J.   174
state function   9
state vector   9, 11, 13, 16
Stern, O.   12, 68, 69, 177
Stern–Gerlach experiment   68, 177
sudden change in the Hamiltonian   141
summation of vectors   7, 25, 39
superconductivity   136
superposition principle   10, 169
symmetric states   95, 96

symmetry   31, 35, 40, 42, 49, 55, 60,
    65–67, 79, 95, 96, 133

target qubit   159
Tegmark, M.   171, 184
teleportation   IX, 154, 156
thermal equilibrium of gas and
    radiation   149
thought experiment   12
time principle   137
time-evolution of the state vector   168
trajectory   6, 19
transition probability   139, 144
    per unit of time   143
translation   40
transmission coefficient   53, 54, 57
Tsui, D.   112
tunnel effect   45, 47, 53, 55
two-electron states   98
two-qubit
    gate   159, 160, 164, 165
    system   159
two-spin states   92, 98, 162

Uhlenbeck, G.   68, 177
unbound problems, one-dimension   51
uncertainty   2, 16
uncertainty principle   17–19, 35, 45,
    109, 181
uncertainty relation, time–energy   140,
    144, 148, 150, 182
unfilled shell   100

unit operator   21, 69
unitary
    matrix   28
    transformation   27, 30, 40, 65
Universitetets Institut for Teoretisk
    Fysik   174, 176
unphysical operators and states   132

valence band   105
validity of quantum mechanics   154,
    184
variational procedure   120–122, 124
vibrational motion   31
virial theorem   34, 89
von Klitzing, K.   111
von Laue, M.   174
von Neumann, J.   170

wave formalism   2
wave function   39, 50–52, 54, 55, 57, 59
wave number   47
wave–particle duality   181
wave-particle duality   15
Wheeler, J.A.   171, 184
Wigner coefficients   71, 75
Woods–Saxon potential   101, 102

Young's double slit experiments   15

Zeeman effect   68, 92, 93
    anomalous   92, 177
zero-point energy   31, 80

Advanced Texts in Physics

This program of advanced texts covers a broad spectrum of topics which are of current and emerging interest in physics. Each book provides a comprehensive and yet accessible introduction to a field at the forefront of modern research. As such, these texts are intended for senior undergraduate and graduate students at the MS and PhD level; however, research scientists seeking an introduction to particular areas of physics will also benefit from the titles in this collection.

Printing and Binding: Strauss GmbH, Mörlenbach